Measuring the Water Status of Plants and Soils

Measuring the Water Status of Plants and Soils

John S. Boyer
College of Marine Studies and College of Agriculture
University of Delaware
Lewes, Delaware

Academic Press
San Diego New York Boston
London Sydney Tokyo Toronto

This book is printed on acid-free paper. ♾

Academic Press, Inc.
A Division of Harcourt Brace & Company
525 B Street, Suite 1900, San Diego, California 92101-4495

United Kingdom Edition published by
Academic Press Limited
24-28 Oval Road, London NW1 7DX

Library of Congress Cataloging-in-Publication Data

Boyer, John S. (John Strickland), date.
Measuring the water status of plants and soils / John S. Boyer,
p. cm.
Companion to: Water relations of plants and soils / Paul J. Kramer and John S. Boyer
Includes bibliographical references (p.) and index.
ISBN 0-12-122260-8
1. Plant-water relationships--Laboratory manuals. 2. Plants. Effect of soil moisture on--Laboratory manuals. I. Kramer, Paul Jackson, date. Water relations of plants and soils. II. Title.
QK870.B68 1995
581.1--dc20 95-10872
CIP

PRINTED IN THE UNITED STATES OF AMERICA
95 96 97 98 99 00 BC 9 8 7 6 5 4 3 2 1

Dedicated to:
Dr. Edward B. Knipling for his contributions and devotion to the development of isopiestic psychrometry,
Dr. Henry C. De Roo for his encouragement and willingness to try new methods,
and my wife Jean whose cheer and understanding kept me writing when I would have liked to have gone sailing.

Contents

Preface

This book is written as a companion to the text by Kramer and Boyer (1995) "Water Relations of Plants and Soils" and is intended for students who need to use some of the methods described there. Water is pervasive in biology, and a student of plants often must face measuring plant water status early in his or her career and virtually alone. Universities sometimes cannot afford to teach a course on the subject but the methods generally are not intuitive. The student must proceed as well as possible, often without a physics or physical chemistry background and with mathematics that have become a little rusty.

For these students and anyone else who wonders how water affects plant growth, I hope the information presented here will be an easy introduction to the measurement techniques. The book is not a detailed review of the literature nor of theory. It does not deal with all the methods for measuring the water status of plants and soils. Instead, it considers the three most used and useful methods and aims at practical laboratory concepts, with considerable effort to keep the mathematical and physical treatments simple and illustrated with examples. Where possible, pictures are employed to give a better understanding of the procedures. I hope my colleagues will forgive the sometimes informal approach and occasional oversimplification.

With this book and an instrument on which to practice, it should be possible to make measurements in plants and soils without some of the pitfalls that are so common. Practically all that is known about plant water relations comes from thoughtful and careful measurements, often by two or more different techniques, and the avoidance of pitfalls may help to approach this ideal.

For the production of this book, grateful acknowledgement is extended to several people. Special thanks are given to Peggy Conlon for typing, editing, and handling the references, and to Dr. An-Ching Tang for the artwork. Thanks are due Karen Lauer and Dr. Michael Lauer for reading parts of the work and for many valuable suggestions. I am immensely grateful for the help and encouragement of Professor Paul J. Kramer in whose laboratory some of the methods were developed while I was a student. I also am indebted to Professor Dr. Ernst Steudle for providing helpful suggestions on preliminary versions of two chapters. Gratitude is extended to Dr. Barry Osmond of the Australian National University for originally suggesting this project and to the Research School of Biological Sciences at the Australian National University and CSIRO of Australia Division of Plant Industry for partial support during later stages of the work.

John S. Boyer

Chapter 1

Why Measure the Water Status of Plants and Soils?

Plant tissues contain large amounts of water, and even larger amounts must be supplied to replace the water lost in transpiration. A maize plant weighing about 800 g at tasseling contains about 700 g of water, and one must supply an additional 20,000 to 50,000 g to grow the plant to this stage. Such a large involvement of water makes it essential to understand how water is used in plant growth, especially since water is the largest input in agriculture. Irrigation has been practiced in agriculture for more than 5000 years and civilizations have fallen because of long dry spells or the failure of irrigation systems.

Historically, water management depended on measuring the time since the last rain or irrigation or occasionally on the extent of wilt of the leaves. These methods had the advantage of simplicity but were too crude to detect early losses in growth. Scientific investigations employed similar methods but were hampered by the difficulty in repeating conditions.

Part of the problem is that plants vary in their response to water. Those with deep roots may prosper when shallow-rooted individuals fail to grow. Early flowering or high water storage can fit some species for a desert existence that others cannot tolerate. Measuring the time since the last rain or the loss in soil water does not take these plant characters into account.

A better approach is to measure the water status of the plant. The differences in water use between species are then included in the measurements, and the varying effects of rainfall and evaporation are integrated as well. There is an increased predictability of plant performance and, for scientific purposes, experimental conditions are more easily measured and reproduced.

Methods of measuring plant water status can generally be classified in three categories. Those in the first category rely on concentrated solutions (osmotica) that cause water release from the tissue. Placing roots in a series of osmotica can indicate which solution causes no water loss or gain, or which causes no change in tissue dimensions. The water status is then expressed in terms of the solution properties causing no change. While these are relatively simple methods, they suffer from the possibility of solute exchange with the tissue. If the membranes of the cells do not completely exclude the solute, the osmotic effectiveness of the solution is less than expected from the concentration and can differ in various tissues (Kramer and Boyer, 1995; Slatyer, 1967; Steudle, 1989). Also, tissue can release water and solute that can change conditions in the osmoticum (Knipling, 1967). Therefore, these methods are not often used.

The second category of water status methods is based on measures of plant water content that are informative when compared to other tissue properties. Typically, the water content is compared to the tissue dry weight or is expressed as a percentage of the maximum water the tissue can hold. The comparison gives a biological baseline or reference which is particularly valuable for determining whether sufficient water has been lost to alter enzyme activities or to concentrate cell constituents.

The third category is based on thermodynamic methods that determine the chemical potential of water in the tissue. These methods have the advantage that the water status is compared to a physically defined reference rather than a biological one, and the physical reference allows the chemical potential to be precisely reproduced at any time or place. The chemical potential has the further advantage that the forces moving water through the soil and plant can be measured.

The latter two categories of methods are the focus of this book, and emphasis is given to the last one because of the wealth of information that can be obtained and the large number of applications that can be made. Many scientific studies now employ measurements of the chemical potential or one of its components, and understanding the methods and their pitfalls is essential for joining this effort.

The development of thermodynamically based methods was given important impetus when Slatyer and Taylor (1960) suggested a new terminology for the water status of plants and soils. This was not the first such suggestion (e.g., see Kramer, 1985), but it was the first to use classical thermodynamics expressed in units already in frequent use.

It included terminology for all the forces in plants and soils, and the ideas were rapidly accepted.

The new terminology accelerated the development of new methods for measuring these forces. A pressure chamber was proposed to measure the tension in the xylem and apoplast of plants (Scholander *et al.*, 1965), and the equipment was used with a vapor pressure osmometer to measure the water potential of leaves soon afterward (Boyer, 1967a). Other methods employed vapor pressures (Monteith and Owen, 1958; Richards and Ogata, 1958; Spanner, 1951) to measure the water potential and were simplified and made more accurate (Boyer and Knipling, 1965). A microcapillary method was developed for directly measuring the turgor inside individual cells (Hüsken *et al.*, 1978). Each had the ability to indicate not only the water status of various parts of plants and soils but also the forces used to move water from place to place.

With the new methods, efforts have been increasingly directed to determining how water moves through the soil-plant system and how metabolism is affected. They show that water is directed to various parts of the soil-plant system according to the difference in water status between the parts (Boyer, 1985; Passioura, 1988). Cell water status often determines the rate of enlargement of the parts and thus is fundamental to the growth process. The forces moving water usually do not directly affect the activity of enzymes because other factors begin to alter enzyme activity before the water status becomes low enough to exert a direct effect (Kramer and Boyer, 1995). The important factors are the availability of products of photosynthesis, the movement of small regulatory molecules, changes in plant growth regulators, and differences in gene expression. As a result of this work, the control of plant metabolism with limited water is seen increasingly as a chemical problem that can be manipulated by altering the availability of particular regulators of enzyme activity and synthesis.

Questions that are now attracting attention include: What conditions lead to changes in regulatory behavior at the molecular level? Is decreased transport of regulatory molecules from the soil a result of the decreased water in the soil or a property of the plant? What signal causes changes in levels of plant growth regulators and gene expression in plant tissue? Answers to these questions will continue to require repeatable water status measurements. For an expanded treatment of these questions, the reader is directed to Kramer and Boyer (1995) and to a symposium volume edited by Close and Bray (1993).

A Little Thermodynamics

Let us think about how the later physically based methods originated. When we consider molecules of any kind, all of them contain energy in their atoms and chemical bonds that can be exchanged with the surroundings by their motions, chemical reactions, and radiational exchanges (here we assume that the isotopic composition remains stable). The energy exchanges always result in a rearrangement of chemical or atomic structure that in itself requires energy (Fig. 1.1). Thus, a fraction of the energy goes to the rearrangements and a fraction to the surroundings, and the latter fraction can be made to do work. The rearrangement energy is the entropy and the energy available for work is the free energy (Fig. 1.1).

It readily can be seen that the amount of work is determined by the number of molecules exchanging energy. Doubling the number of molecules doubles the work, all other factors remaining constant. Often, however, it is more desirable to know the work per molecule or per mole of molecules than the total work. J. Willard Gibbs (1931) recognized this and defined the term "potential" and symbol μ as the way to describe the work that a mole of molecules can do.

The work is not known in absolute terms because the total amount of energy in molecules is not known. Therefore, the work is determined by comparing the chemical potential of the system with a reference potential. For liquid water, the reference has been chosen to be pure unrestrained water at atmospheric pressure, a defined gravitational position, and the same temperature as the system being compared. If we define the chemical potential of the system to be measured as μ_w and the chemical potential of the reference as μ_o, $(\mu_w - \mu_o)$ is the comparison we wish to make. When the system is not pure water, the μ_w is lower than μ_o, and $(\mu_w - \mu_o)$ is negative. When the system is pure water, $(\mu_w - \mu_o)$ is zero. When the system is varied in pressure or gravitational position in a water column, $(\mu_w - \mu_o)$ can be positive or negative.

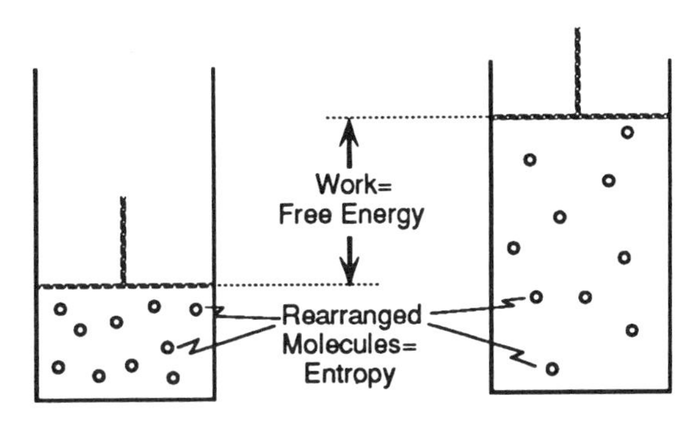

Figure 1.1. Molecular changes occurring when work is done by a gas and piston. The rise in the piston represents work which is the force of the piston times the distance moved. On the left, the molecules are close together and on the right they are far apart. The work cannot occur unless this molecular rearrangement takes place. The energy consumed in the rearrangement is the entropy. The remaining energy raises the piston, does the work, and is the free energy.

The $(\mu_w - \mu_o)$ is the energy state of the molecules. It does not matter how the molecules get to that state, the energy is the same whenever $(\mu_w - \mu_o)$ is at the same level. The energy represents the maximum work that can be done if the molecules are part of an ideal machine. Pure water moving through a selective membrane into a solution on the other side is a machine allowing work to be done as molecules on one side escape from the bulk and pass through the membrane to the other side. If the membrane allows water to pass but not solute (the membrane reflects solute), more water will move to the solution side than to the other side because the free energy of the pure water is higher than in the solution. The work is determined by the potential difference on the two sides of the membrane and the net volume of water moved. The work can be measured by opposing this movement with a chemical potential that counters the movement.

If the membrane is not reflective for solute, the volume of water moving into the solution is the same as the volume of water and solute moving in the opposite direction. No work is done because there is no net volume change. Nevertheless, at the beginning, the $(\mu_w - \mu_o)$ is the same as when the reflective membrane was present. Thus, the ability to do work is identical but the work actually done depends on the characteristics of the machine.

This example illustrates that the $(\mu_w - \mu_o)$ is an intrinsic property of the molecules. The membrane simply determines the work extracted from the molecules. The reflectiveness of the membrane is usually described by the reflection coefficient which is 1 for a perfectly reflective membrane but 0 for a nonreflective one. This is important for anyone studying water movement in plants and soils. The osmotic effectiveness of a solution is determined by the membrane reflectiveness from the beginning even though large concentration differences exist on the two sides of the membrane. This is one reason why methods of measuring water status with osmotica may not give accurate data unless the membranes have a reflection coefficient of 1.

The idea of Slatyer and Taylor (1960) was to express the chemical potential in pressure units to make it simpler to apply to plant and soil systems. This was done by dividing $(\mu_w - \mu_o)$ by the partial molal volume of liquid water $\bar{V}_w$ to give the water potential Ψ_w:

$$\Psi_w = \frac{(\mu_w - \mu_o)}{\bar{V}_w}. \tag{1.1}$$

Because the units for $(\mu_w - \mu_o)$ are energy per mol and for $\bar{V}_w$ are volume per mol, the units of Ψ_w are energy per volume = force per unit area = pressure. The pressure is usually expressed in megapascals (MPa) where 1 megapascal = 10^6 pascals = 10^6 newtons·m^{-2} = 1 joule·m^{-3} = 10 bars = 9.87 atmospheres or 145 pounds per square inch.

The $\bar{V}_w$ is the volume of a mole of liquid water mixed with other molecules in the system and is nearly a constant 18 cm^3·mol^{-1} over most of the temperatures and water contents of cells and soils. Therefore, the Ψ_w is simply $(\mu_w - \mu_o)$ divided by a constant. In concentrated solutions, dry soils, and other systems of low water content, this simplification may not hold because interactions between water and the other molecules can be so extensive that 1 mol of water no longer occupies 18 cm^3. In this case, the proportionality breaks down, and $(\mu_w - \mu_o)$ should be used whenever Ψ_w is below about -10 MPa.

The chemical potential can be measured from the relationship between pressure and volume. In Fig. 1.1, work is done by the expansion of the gas against the piston. The work is the distance the piston moves times the force exerted by the piston, which is described by $dV \cdot P$ and has units of m^3 x force$\cdot m^{-2}$ = force$\cdot$distance. The work can be measured by holding the volume constant and measuring the change in pressure or by holding the pressure constant and measuring the change in volume. Of the two the former is easier so that at constant temperature

$$\begin{aligned}(\mu_w - \mu_o) &= \int dV \cdot P = \bar{V}_w \int_0^P dP \\ &= \bar{V}_w \cdot (P - 0),\end{aligned} \tag{1.2}$$

where the volume is held constant. Notice that, because the pressure is in the *liquid* water whose volume is essentially incompressible, the volume of 1 mol of water is a constant $\bar{V}_w$. The constant does not enter the integration and the equation gives the maximum work that 1 mol of liquid water molecules can do. From Eq. 1.2, the water potential is

$$\Psi_w = \frac{(\mu_w - \mu_o)}{\bar{V}_w} = P, \tag{1.3}$$

which indicates that the water potential is equivalent to a pressure, usually negative. To measure this pressure, we create a counteracting pressure that prevents work, i.e., prevents water from moving in the plant or soil system. The measuring pressure equals P. Because the measuring pressure counteracts P, it is an equilibrium measurement.

Measuring work with vapor pressures follows a similar procedure. The pressures are applied to water vapor in the *gas* phase. In this case, the volume of a mole of water is no longer constant. From the gas law, the volume of a mole of gas molecules is $v = RT/e$ where we use lower case v and e to indicate the volume and pressure in the gas phase. The chemical potential in the gas phase is

$$(\mu_w - \mu_o) = \int dv \cdot e = RT \int_{e_o}^{e_w} \frac{de}{e} \qquad (1.4)$$

$$= RT \ln \frac{e_w}{e_o}.$$

Here, R is the gas constant (8.3143×10^{-6} $m^3 \cdot MPa \cdot mol^{-1} \cdot K^{-1}$), and T is the Kelvin temperature (K), which is held constant. Therefore, RT does not enter the integration. The water potential is

$$\Psi_w = \frac{RT}{\bar{V}_w} \ln \frac{e_w}{e_o}, \qquad (1.5)$$

which expresses the water potential in the usual way by dividing the chemical potential by the partial molal volume of *liquid* water.

One may visualize that just as the chemical potential affects the ability of liquid water molecules to escape through a membrane, it will affect the ability of liquid water molecules to escape into the gas phase (evaporate). If we can measure the ability to evaporate, we have a measure of the chemical potential in the liquid. To measure the ability to evaporate, we need only to create a partial pressure for water in the gas that matches the vapor pressure of water in the liquid, preventing evaporation. This is the equilibrium vapor pressure, and the Ψ_w in the gas equals the Ψ_w in the liquid. Equation 1.4 tells us that $(\mu_w - \mu_o)$ of the gas is related to the ability to evaporate according to the ratio of the vapor pressure of the system (e_w) to the vapor pressure of the reference (e_o), i.e., the relative humidity at the temperature of the system.

Of course, temperature has large effects on the vapor pressure of water but Eq. 1.5 compares e_w and e_o at the same temperature and e_w/e_o responds only to nonthermal effects (concentration, pressure, and so on). Temperature has its effect mostly on T (which decreases at lower temperatures) and slightly on $\bar{V}_w$ (which decreases, then increases at lower temperatures). As T decreases to absolute zero, Ψ_w approaches zero. Similarly, P in Eq. 1.3 shows no thermal response because of the isothermal nature of the measurement, but it will respond to the change in potential according to T in the measured system. Appendix 3.2 is a practical demonstration of these facts as the osmotic potentials of sucrose solutions become less negative when the Kelvin temperatures decrease. As a consequence, the cell changes slightly in potential as T varies, and the e_w or P used to measure the potential will vary accordingly. The response is not large because there is only a narrow range of Kelvin temperatures in which biological systems exist, and the e_w or P measurements respond similarly and predictably.

This book will treat pressure and vapor pressure methods of measuring plant water status. The pressure chamber and pressure probe use pressures to measure Ψ_w and thus Eqs. 1.2 and 1.3 apply (Chaps. 2 and 4). The thermocouple psychrometer uses vapor pressures to measure Ψ_w and Eqs. 1.4 and 1.5 apply (Chap. 3). The equal signs in the equations indicate that the measurements are made at equilibrium, termed thermodynamic equilibrium.

The Value of Thermodynamic Equilibrium

Thermodynamics tells us that it is simplest to measure the energy state of molecules by using a system that prevents any work from being done, that is, by preventing the molecules from changing to a lower energy state during the measurement. In practical terms, this is done by using a measuring system to counterbalance the tendency of the molecules to do work. In this case, the measuring system is in thermodynamic equilibrium with the molecules being measured. For pressure in liquids, we counterbalance the pressure with an opposing but equal pressure. For vapor pressures, we create a vapor pressure that equals the vapor pressure of the molecules, preventing evaporation and again counterbalancing the pressure being measured.

Thermodynamic equilibrium is valuable first because it allows molecular energies to be determined without changing the molecules. If measurements are not at equilibrium, the measured molecules change energy and the measurement is affected by all the factors that affect energy change: the size of the energy differences that drive the process, the resistances to energy change imposed by the apparatus, the position of the exchanging molecules relative to each other, and so on. Measuring pressure without using a counterbalancing pressure on liquids, for example, allows flow to occur. While the flow rate can give

Figure 1.2. Accurate measurements of the pressure in this pore are easiest when there is no flow!

information about the pressure, one also must know the pressure difference between the measured molecules and the measuring instrument, the resistance to flow between the two systems, and many other factors. The pressure measuring instrument can be calibrated if conditions can be precisely controlled and repeated, but additional complexity is added. Therefore, it is preferable to measure at equilibrium where there is no flow.

The second reason thermodynamic equilibrium is important is that energy standards are readily available. Reference pressures are precisely known for the atmosphere or at the base of a column of water. Vapor pressures of solutions are well known and standards are readily available in the laboratory. As a result, equilibrium methods need little if any calibration, which is a great simplification. Moreover, because the factors affecting energy exchange between the molecules and measured system do not affect the measurement, determinations have less variability (see Chap. 5, Fig. 5.1 for an example). All in all, more accurate measurements are the result (Fig. 1.2).

Additional Readings

For readers interested in pursuing these concepts further, a number of papers and reviews treat various aspects of measuring plant and soil water status. Particularly recommended are papers by Slatyer and Taylor (1960), Ritchie and Hinckley (1971), Tyree and Hammel (1972), and Brown and Oosterhuis (1992), reviews by Boyer (1969b) and Zimmermann and Steudle (1978), and the symposium proceedings published by Brown and Van Haveren (1972) and by Hanks and Brown (1987). A review by Barrs (1968) is useful for a historical description of older methods but contains several errors concerning more recent methods.

Chapter 2

Pressure Chamber

Pressure chambers are the most widely used field instruments for measuring plant water status. They are portable, allow rapid measurements, and are sturdy. Temperature needs little control and no complex instrumentation is required. The tissue simply is sealed into the top of the chamber in such a way that most is inside and only a small amount extends outside through the top (Fig. 2.1A). The seal gives an airtight barrier between the interior and the atmospheric pressure outside (Fig. 2.1B, C). This allows the tissue to be pressurized inside, forcing water toward the outside. The pressure necessary to hold the water at the outside surface measures the water status of the tissue. The more dehydrated the tissue, the more pressure is required.

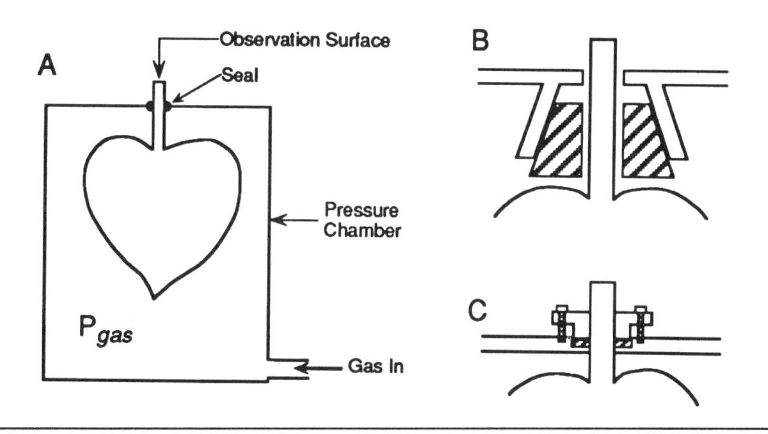

Figure 2.1. Pressure chamber design. A) Basic layout of chamber and tissue during pressurization. B) Enlarged view of a seal that is tightened by gas pressure; the seal around the tissue becomes tighter as pressure drives the rubber stopper (cross-hatched) deeper into the cone below the chamber top. C) Enlarged view of a seal that is tightened manually; tightening the screws forces the rubber seal against the tissue.

The method has been used successfully with leaves, branches, and roots and can provide information about all of the components of the water potential (Scholander *et al.*, 1965; Tyree and Hammel, 1972). Because the chamber is portable, it can be moved to the experimental site, and conditions in the plant can be left undisturbed until the moment of sampling. However, because plants use large pressures to move water, large pressures are needed during measurements. Therefore, pressure chambers must be built strongly and the seal must hold the tissue in place against large forces. Early efforts with pressure chambers were plagued by explosions that prevented the method from being developed (Dixon, 1914). After Scholander and his colleagues built a safer unit (Scholander *et al.*, 1964; 1965), the method became better understood and Boyer (1967a) showed that it could be used to measure the water potential. Commercial versions are now available (Appendix 2.1).

Principles of the Method

The method is based on the concept that the water potential in cells creates a tension (negative pressure) in the cell walls that pulls water toward the cells from the xylem, the root cells, and finally the soil (Fig. 2.2A). Excising a plant part causes the xylem water to pull back into the xylem (Fig. 2.2B). Applying pressure to the tissue raises the cell water potential and forces water out and into the xylem which is open to the atmosphere outside of the chamber. Xylem solution eventually appears at this surface when the applied pressure fully opposes the tension originally in the sample (Fig. 2.2C). Scholander *et al.* (1965) considered the pressure to be a direct measure of the tension in the xylem because of the continuous liquid phase extending into the cell walls.

The liquid moving in the walls and xylem is not pure water. Roots absorb salts from the soil and deliver them to the shoot via the xylem. Together with certain organic constituents traveling in the xylem, the xylem and cell wall solution contain sufficient solute to have osmotic potentials as low as -0.4 MPa (Boyer, 1967a). Boyer (1967a) showed that if solute effects were taken into account, the pressure chamber could be used to measure the tissue water potential. Since then, the pressure chamber has been widely used for measuring tissue water potentials.

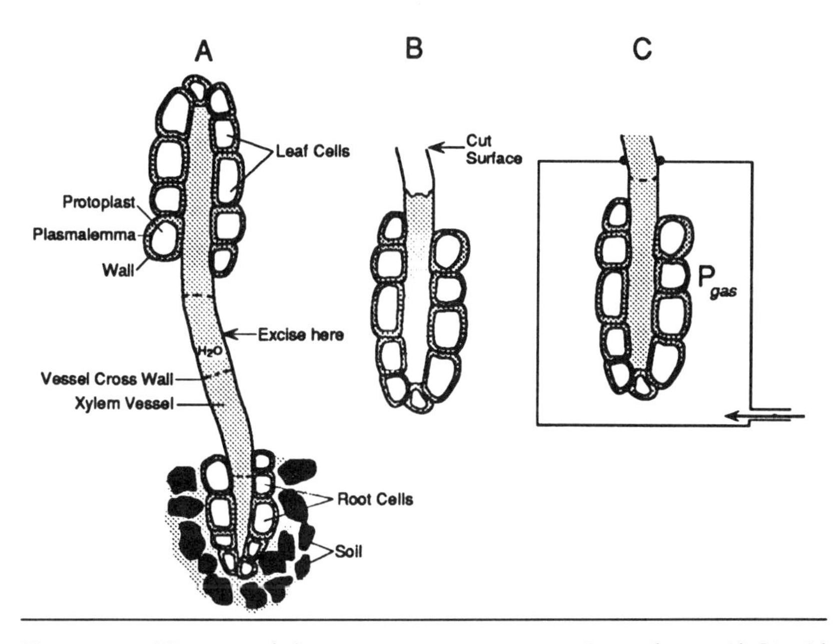

Figure 2.2. Diagram of the water transport system in a plant. A) Liquid continuity occurs between the soil solution and the cells inside the leaf. The cells (protoplasts, open spaces) are water-filled and bounded by a membrane. The cell walls and xylem (apoplast, stippled spaces) are also water-filled. The water in the apoplast is continuous with water in the soil except that in many instances there are waxy substances (Casparian strips, not shown) in the walls of the root cells that force water to flow through the root protoplasts. Leaf protoplasts transmit low water potentials as tensions to the apoplast and the soil. B) Excising a leaf opens the xylem to the atmosphere. The xylem solution retracts to cross walls in the vessels where sufficiently small pores exist to prevent solution from retracting farther. C) Mounting the leaf in a pressure chamber allows pressure to be applied that returns the xylem solution to its position in the intact plant in A. The pressure P_{gas} counteracts the tension (negative pressure) exerted on the xylem/apoplast solution in the intact plant and thus is a measure of the tension.

Pressure Chamber Theory

These concepts can be formalized by considering the components that contribute to the water potential. In cells and tissues, the major ones are solute, pressure, solids (porous solids also termed matrices), and gravity. The distinction between them sometimes becomes blurred because of the difficulty in categorizing the forces, especially those arising from solutes and solids. In cells, macromolecules can dissolve or sometimes precipitate to form a gel (porous matrix) or form aggregates that may or may not be in solution. Despite this complexity, we will consider aggregates to be in solution unless they precipitate, a practice also followed by J. Willard Gibbs (1931). We will distinguish between pressure generated in pores and pressure applied externally (Gibbs, 1931). Also, as discussed in Chap. 1, we will describe the forces on a unit area basis (pressure) which is proportional to the free energy per mole of molecules (see Chap. 1). Accordingly, the components of the water potential are

$$\Psi_w = \Psi_s + \Psi_p + \Psi_m + \Psi_g, \tag{2.1}$$

where the subscripts s, p, m, and g represent the effects of solute, pressure, matrix, and gravity, respectively, and Ψ_w is the water potential. Each potential refers to the same point in the solution, and each component is additive algebraically according to whether it increases (positive) or decreases (negative) the Ψ_w at that point. The increase or decrease is always relative to pure water at atmospheric pressure, at the same temperature as the solution (see Chap. 1).

The components affect Ψ_w in specific ways. Solute lowers the chemical potential of water by reducing the tendency of water molecules to escape from each other compared to pure water because some of the solution volume is occupied by solute molecules that dilute the water molecules, decreasing the number able to escape. In a similar fashion, porous solids that are wettable occupy volume and cause surface effects that reduce the escaping tendency of the water in the matrix. External pressures applied to the liquid increase the escaping tendency of the water if they are above atmospheric but reduce the escaping tendency if they are below atmospheric. Gravity affects pressures because of the weight of the water, and the escaping tendency of water is increased or decreased depending on whether gravity increases or decreases the local pressure relative to atmospheric pressure. Pressures are high at the bottom of the ocean for this reason but low at the top of a tall tree. Each component decreases Ψ_w (is negative in Eq. 2.1) except pressure and gravity above the atmospheric level, which increase Ψ_w (are positive in Eq. 2.1).

For most of our purposes, gravitational potentials will be ignored because they become significant only at heights greater than 1 meter in vertical water columns. In these cases, Eq. 2.1 reduces to:

$$\Psi_w = \Psi_s + \Psi_p + \Psi_m. \tag{2.2}$$

Water in plants generally has a negative Ψ_w because Ψ_s and Ψ_m are negative and Ψ_p does not fully compensate for them. Water will move toward more negative Ψ_w or more negative components of Ψ_w, and plants use this principle to extract water from the soil.

We may further conceptualize plant water by recognizing that it is located in two compartments separated by a differentially permeable membrane (Fig. 2.3). The first compartment is the interior of the cells (the protoplasts which collectively are the symplast) and the second is the cell walls and xylem outside of the protoplasts (collectively the apoplast). The membrane separating the compartments is the plasmalemma of each cell, and it allows water to move freely but little solute (i.e., the membrane is differentially permeable and reflects solutes).

Figure 2.4 shows that in the protoplast compartment (Fig. 2.4A), there is a concentrated solution ($\Psi_{s(p)}$) and usually a pressure above atmospheric (the turgor, $\Psi_{p(p)}$) so that the water potential ($\Psi_{w(p)}$) is

$$\Psi_{w(p)} = \Psi_{s(p)} + \Psi_{p(p)}, \tag{2.3}$$

where the subscript *(p)* denotes the protoplast compartment. The matric potential is generally negligible in the protoplast compartment ($\Psi_{m(p)} = 0$; Boyer, 1967b).

In the apoplast compartment (Fig. 2.4B), there is a dilute solution ($\Psi_{s(a)}$) and no turgor. Instead, there are surfaces arising from the porous matrix of the cell walls (Fig. 2.3, inset) and these generate a matric potential $\Psi_{m(a)}$ which is expressed mostly as a tension, i.e., negative pressure when the pores are water-filled. These components are expressed by

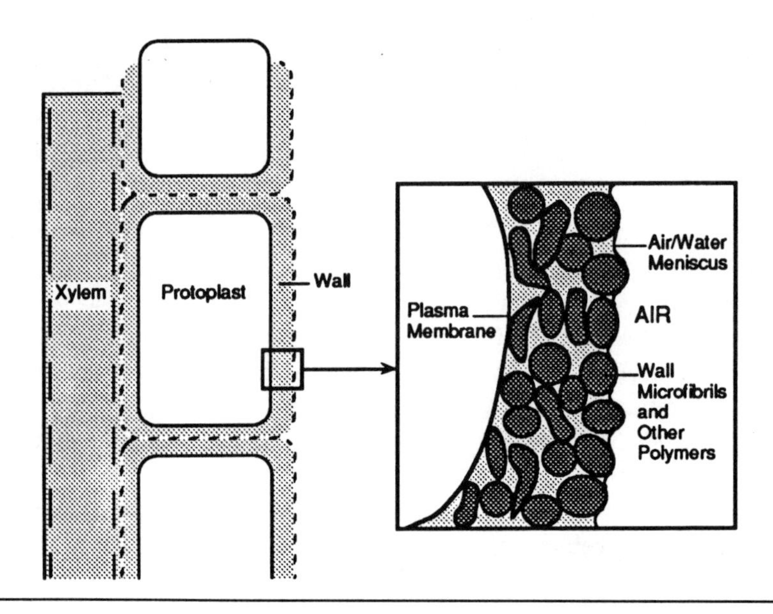

Figure 2.3. Enlarged view of compartmentation in plant tissues. The first compartment is inside the cells (protoplast, open space) and the second is in the cell walls and xylem (apoplast, stippled spaces). The two compartments are separated by the plasmalemma (plasma membrane in magnified inset). The protoplasts are water-filled. The walls contain water in the pores held by the hydrophilic surfaces of the pores and by surface tension at each air/water meniscus. The wall pores are so small that they withstand high tensions without draining.

$$\Psi_{w(a)} = \Psi_{s(a)} + \Psi_{m(a)}, \tag{2.4}$$

where the subscript *(a)* denotes the apoplast compartment.

The water potential in each protoplast is almost always the same as in its own cell wall (Molz and Ferrier, 1982) as shown in Fig. 2.4C:

$$\Psi_{w(a)} = \Psi_{w(p)}. \tag{2.5}$$

Substituting Eq. 2.3 and 2.4 in Eq. 2.5 gives

$$\Psi_{s(a)} + \Psi_{m(a)} = \Psi_{s(p)} + \Psi_{p(p)}, \tag{2.6}$$

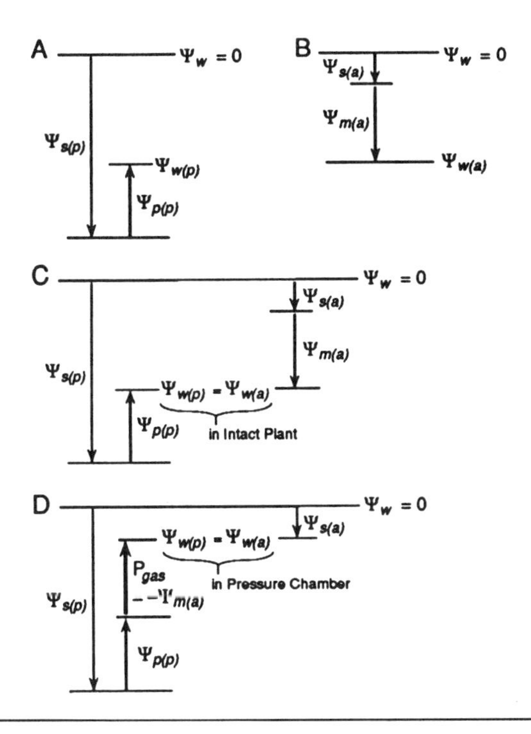

Figure 2.4. Potential diagrams showing water potential and its components inside cells (A) and in walls and xylem (B). The direction of the arrow indicates whether the potential is raised or lowered by each component. The cell interior A is essentially in equilibrium with its wall B so that the diagrams in A and B can be equated as in (C). Pressure (P_{gas}) applied to tissue in the pressure chamber raises the potential of the cell interior as in (D). Water moves out and hydrates the wall, raising its potential. When P_{gas} balances $\Psi_{m(a)}$ as shown in D, liquid appears at the cut surface and does not move at equilibrium.

which shows that the components of the water potential in the protoplasts are different from those in the apoplast but they balance each other locally. Note that the turgor in the cells is positive ($\Psi_{p(p)}$) and the water in the apoplast is under tension ($\Psi_{m(a)}$). This causes a large pressure difference across the plasmalemma. Were it not for the constraining effect of the cell wall, the plasmalemma would burst.

Upon pressurization in a pressure chamber, as in Fig. 2.2C, the water in the cells is uniformly exposed to an external pressure in addition to the turgor (Fig. 2.4D). This raises the cell water potential above that of the xylem, and water flows into the xylem. By adjusting the pressure, the flow can be stopped when water just fills the xylem. At this balancing pressure P_{gas}, the water has returned to its original position in the intact plant where it forms a stationary flat film without any excess on the cut surface (Fig. 2.2C). This pressure exactly relieves the tension that had been acting on the xylem solution (Fig. 2.4D). The negative of P_{gas} thus measures the tension in the apoplast, i.e., the matric potential of the apoplast:

$$-P_{gas} = \Psi_{m(a)}. \qquad (2.7)$$

Substituting Eq. 2.7 in Eq. 2.4, it can be seen that the water potential of the apoplast is the sum of $-P_{gas}$ and the osmotic potential of the apoplast solution $\Psi_{s(a)}$. From Eq. 2.5,

$$\Psi_{s(a)} - P_{gas} = \Psi_{w(a)} = \Psi_{w(p)}, \qquad (2.8)$$

and the water potential of the tissue is determined from $\Psi_{s(a)} - P_{gas}$ (Boyer, 1967a). The $\Psi_{s(a)}$ is measured by overpressuring the tissue, collecting a small amount of exudate from the xylem, and determining its osmotic potential in an osmometer (see Chap. 3).

Three basic principles are demonstrated by these relationships. First, the pressure chamber measures the tension in the xylem and cell walls because the applied pressure relieves the tension and the xylem is directly observed. Second, the measurements require an equilibrium between the pressure and the xylem solution (hence the equal signs in Eqs. 2.7 and 2.8). One must make the measurements at equilibrium pressures (no water moves in or out of the tissue) to have a valid measurement. Third, the applied pressure raises the water potential in the tissue. During the measurement, the tissue does not have the same potential as in the intact plant (Fig. 2.4D).

SIGNIFICANCE OF THE THEORY

The ability of water to move depends on its water potential or the components of the water potential, specifically the osmotic potential, pressure, or matric potential. In plants, low potentials are frequent because the shoot tissues become dehydrated on a daily basis. The low potentials create a tension on water in the xylem and, when the pull is large enough, water moves into the plant. When the rate of uptake equals the rate of evaporation, the water potential becomes constant. Anything that changes the rate of loss or uptake will alter the water potential of the cells, and the potential can change quickly. Variations in wind speed or brief periods of cloudy weather can cause especially large changes, and taking the pressure chamber to the field is a good way to convince yourself of this.

As a consequence, the pressures in the xylem vary markedly during the day. Before sun up when transpiration is low, pressures often are above atmospheric. The roots absorb salts from the soil and, because they exhibit active mechanisms to transfer salts to the xylem, they create a sufficient concentration in the xylem to attract water osmotically from the soil (Fig. 2.2). Pressures build up in the xylem just as they do in cells and are termed root pressures. Root pressures of 0.05 to 0.4 MPa are frequent in rapidly growing plants, and occasionally they are even higher (to 0.6 MPa). The pressure chamber cannot measure these positive pressures or the osmotic potential in the xylem.

After sun up, transpiration begins and the pressure falls in the xylem, frequently reaching tensions of -1 to -2 MPA as the cells dehydrate. Water is pulled from the soil much as a wick pulls liquid from its surroundings, except that cells are embedded in the wick and the tensions are controlled by the cells. Together with the xylem, the wick-like action can move water over large distances very rapidly. The pressure chamber can measure these tensions.

The pores in the cell walls are in contact with the water in the xylem but they do not drain because water molecules adhere to the molecules in the walls and to each other, and are attracted to the solutes adsorbed to the walls. The wall pores have small diameters (about 5 to 8 nm), and surface tension keeps the pores water-filled against large tensions (Figs. 2.2 and 2.3). The xylem vessels have larger internal diameters but they generally remain water-filled unless tensions become more negative than about -1.0 to -2.0 MPa. In that case, xylem water can drain because the water column tends to break (cavitate), but the small pores in the surrounding walls remain water-filled. Breaks in the xylem water can diminish transport in the vascular system.

Pressure applied with the pressure chamber is external and the gas penetrates the tissue through the intercellular air spaces, thus surrounding each cell with uniform pressure (Figs. 2.2 and 2.3). Only the cell to cell contact areas are not in contact with the gas, and liquid can flow from cell to cell through these areas. Thus, the pressure "squeezes" the liquid in the cell toward any region where the cell to cell contact is at a lower potential.

The temperature does not enter any of the above equations because it is uniform in the sample. However, pressure-volume work depends on the Kelvin temperature, and the water potential of the tissue becomes less negative as the temperature decreases within the biological range. You will see this as a slight decrease in the balancing pressure as a leaf sample becomes colder even though there is no change in the water content of the sample. The theoretical basis for the temperature dependence is treated in Kramer and Boyer (1995).

Types of Pressure Chambers

All pressure chambers are constructed similarly except for the seal in the top. Figure 2.1B shows a pressure-activated seal that relies on the slippage of the sealing material, usually rubber, in a cone-shaped cavity built into the underside of the chamber top. As pressure increases inside the chamber, the outward force pushes the sealing material into the cone. The decreasing diameter of the cone tightens the seal around the plant part.

Pressure-driven seals have the advantage that they are quick and automatic. The surface between the seal and the cone needs to be lubricated but otherwise no maintenance is required. The disadvantage is that there is no control over the force applied to the tissue and the seals must be long to align the movement in the cone. The long seal restricts samples to long-stemmed branches and leaves with long petioles. Because the force applied to the tissue can become very large at high pressures, it can damage soft tissues or even interrupt flow. Thus, the design is best for woody stems.

Another type of seal involves a rubber packing gland whose tightness can be adjusted by the operator (Fig. 2.1C). The rubber is enclosed in a well on top of the chamber. A packing plate sits on top of the rubber and can be pushed down. Because the rubber cannot deform outward, it deforms into the center, filling it according to the force on the packing plate. The deformation seals the stem or petiole.

Manually tightened seals have the advantage that a minimum of force is applied to the plant material. The operator listens for escaping gas and tightens just enough to prevent audible leakage. The seal is small so that short-stemmed samples can be used. This type of seal is preferred over the pressure-driven seal because it is less likely to damage the tissue.

Seals can vary in diameter to allow large diameter stems to be used but the stem must be especially secure because pressures exert a force per unit area, and doubling the radius of the sealed tissue increases the force fourfold. In this situation, seals must exert a much larger force on the tissue to hold it in the pressure chamber.

How to Make Measurements

PRELIMINARY CHECKS

The static pressure inside a confining vessel is the same on all the walls, and the pressure gauge may be mounted on the gas feed line rather than on the chamber itself. Before using the instrument for the first time, the gauge may need to be checked for accuracy. Gauges are available at three levels of accuracy: standard, test, and master test. The standard gauge is used on equipment requiring moderately accurate pressure readings. Test gauges are used to calibrate standard gauges and give more reproducible readings. Master test gauges have a calibration traceable to the National Bureau of Standards and are typically used to calibrate test gauges. Of the three, test gauges are preferred for pressure chambers and their accuracy can be assumed. If standard gauges are used, they should be calibrated at least with a test gauge.

Before pressurizing a pressure chamber, test for its ability to withstand high pressures. Measurements with plants usually do not exceed 6 MPa but, whatever the maximum, tests should be at pressures at least twice the maximum. For the test, fill the chamber with water so that there is no air. Seal a metal rod securely in the top in place of the tissue. Pressurize the water and check for leaks. The incompressibility of the water ensures that any failure will not be dangerous.

PROCEDURE

1) After checking that the pressure chamber and seal are in good condition, clean, and dry, check that the incoming gas will enter close to the bottom. Cover the bottom with a layer of water so that the incoming gas passes through the water. Make a baffle to prevent water from splashing onto the tissue (Fig. 2.5). Line the walls with wet filter paper. Connect a cylinder of compressed air to the gas line.

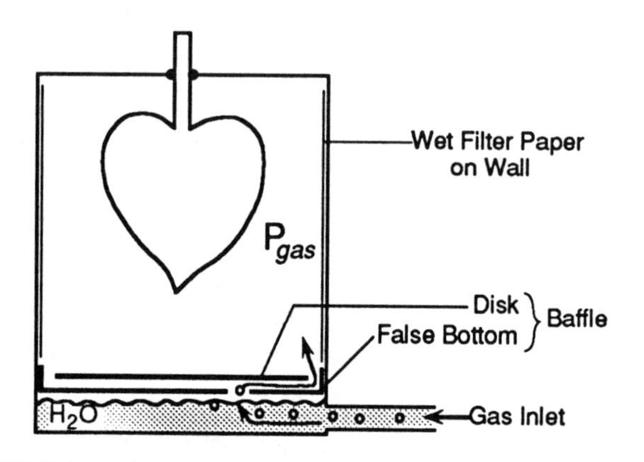

Figure 2.5. Water in the bottom of a pressure chamber reduces overheating of the entering gas and humidifies the air around the leaf. A two-part baffle prevents splashing on the leaf. Normally, the chamber wall is also lined with wet filter paper.

2) Select the sample, avoiding damaged tissue whenever possible. Excise the tissue with a razor blade, insert it swiftly into the seal in the chamber top, and assemble the chamber. The time from excision to sealing the chamber should be no longer than 10 sec to avoid dehydrating the tissue.

If the time is longer than 10 sec, use a humidified glove box to load the sample in the seal and assemble the chamber (see Chap. 3). Work in low light to avoid heating and dehydrating the tissue. Alternately, immediately before excision, enclose the tissue to be pressurized in a flexible plastic bag to retard evaporation (Turner and Long, 1980). Seal the bag enough to inhibit evaporation but allow gas

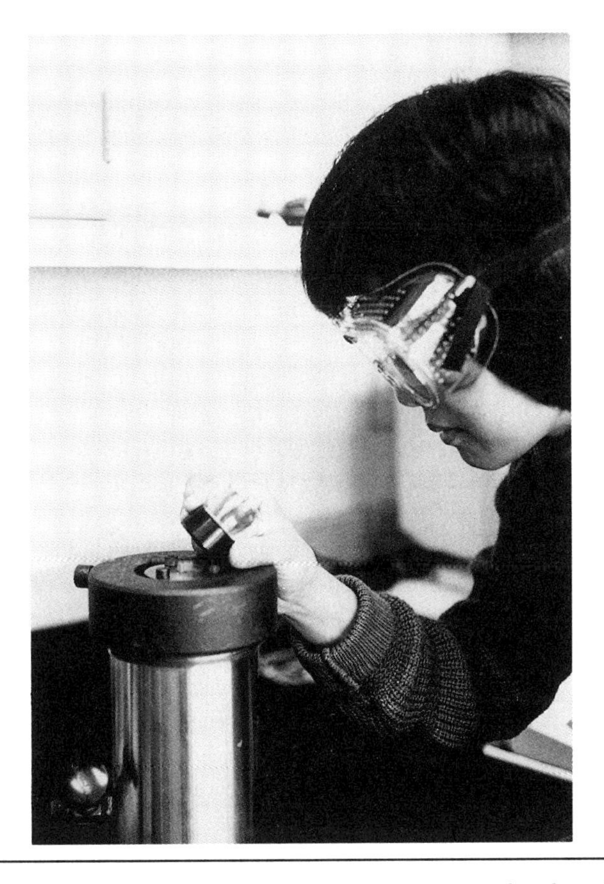

Figure 2.6. Measuring the xylem tension with a pressure chamber. Note that the observer always stays to the side of the apparatus and wears safety glasses in case tissue is blown out of the chamber.

to escape under pressure, shade the bag, excise the sample, and load into the chamber as above.

3) After the tissue has been placed in the chamber, apply a small amount of pressure and check for leaks. If air is leaking through a manually sealed unit, slowly tighten the seal until audible leakage stops. Raise the pressure slowly and in small steps.

4) Observe the cut surface of the tissue as pressure is being applied. ALWAYS OBSERVE FROM THE SIDE RATHER THAN ABOVE THE

CHAMBER IN CASE THE TISSUE IS BLOWN OUT OF THE SEAL (Fig. 2.6). Do not recut the tissue because the initial excision is the reference position marking the location of water in the xylem in the intact plant. It is to this position that the xylem solution must be returned by the pressure. Increase the pressure until liquid is standing on the cut surface.

5) As pressure increases, it is normal to observe gas bubbles on the surface, but they should be forming slowly enough not to obscure the arrival of the liquid. After liquid appears, reduce the pressure and allow the liquid to be pulled into the tissue until a wet film is all that remains on the cut surface. This is the position of the xylem solution before excising the sample, and it should require a balancing pressure that exactly opposes the tension in the xylem before excision. The meniscus is flat indicating that the water is not constrained by tensions that would otherwise be operating. Adjust the pressure so that the water film remains at the cut surface. For your first measurements, satisfy yourself that the liquid film is maintained for 30 to 60 min without changing the final balancing pressure. During this time, evaporation from the film can be prevented by lining a vial with wet filter paper and inverting it over the cut surface. For routine measurements, it will suffice to observe the meniscus for only 1 or 2 min. The balancing pressure is the negative of the tension in the xylem (Eq. 2.7).

6) After determining the balancing pressure, rinse the surface with water, dry the surface, overpressure, and collect a small (10-20 μl) sample of xylem solution in a microliter syringe for a measurement of the xylem osmotic potential. Measure the osmotic potential with a microliter osmometer or psychrometer (see Chap. 3). If the osmotic potential is sufficiently close to zero, this step can be omitted in subsequent determinations.

7) Release the air in the chamber and remove the tissue. Inspect for damage from pressurizing and sealing.

CALCULATING WATER POTENTIALS FROM PRESSURE CHAMBER DATA

Gauge pressures are converted to megapascals (MPa) or bars according to 1 MPa = 10 bars = 10^6 newtons·m^{-2} = 10^7 dynes·cm^{-2} = 145 lb·in^{-2} = 9.87 atmospheres. The apoplast osmotic potential (negative) is calculated according to the osmometer instructions (see Chap. 3). The water potential of the tissue is the sum of the apoplast osmotic potential and the negative of the balancing pressure (Eq. 2.8).

Working with Plant Tissue

There are gradients in water potential in plants. As a result, tissue samples are not uniform, and large samples used with pressure chambers may have significantly different water potentials in different parts. The xylem is the source of water for many plant organs and usually is the wetter part of the gradient. After some time elapses in the pressure chamber, the water in the tissue equilibrates and the chamber indicates an average. Usually, the average is approached within 10 min which is the time required for a pressure chamber measurement. However, in some fleshy samples, times can be much longer (hours or days).

Tyree and Hammel (1972) showed that the average is determined by how much water is present in each part of the sample as well as by the potential of the water (also see Fig. 3.19) according to

$$Average\ \Psi_w = \frac{\Sigma V^i \cdot \Psi_w^i}{V}, \qquad (2.9)$$

where V^i is the water volume in the protoplasm of cell i, Ψ_w^i is the water potential of cell i, and V is the total water volume in the protoplasm of all the cells (the symplasm). The symbol Σ adds the contribution of all the cell V^i and Ψ_w^i in the tissue. Dividing by V gives a volume-weighted average. The volume of water in the apoplast is not considered in this calculation because the xylem and wall matrix are considered to be incompressible (but see section on Changes in Xylem Dimensions).

Equation 2.9 shows that those parts of the tissue containing the largest water volume make the largest contribution to the water potential measured with a pressure chamber. The cells far from the xylem account for a larger volume than the few cells next to the xylem, and the potential of the far cells will dominate the volume average in a pressure chamber. You can observe this by rapidly pressurizing a sample, which often forces a transient show of xylem solution at a low pressure because of the release of water from the wetter cells immediately around the xylem. However, the solution disappears and the final stable reading is always at a higher pressure reflecting the eventual contribution of the far cells to the volume average. The volume-averaging concept applies to the components of Ψ as well as to any other cell parameters depending on Ψ and measured at the tissue level. With care, the volume average measured with the pressure chamber should be the same as the volume average in the intact plant before sampling.

Other factors can change the Ψ_w, and a particular problem for the pressure chamber is the inherently dehydrating nature of the measurement. Only excised tissue can be used, which eliminates water uptake. The gas entering the pressure chamber is dry and warm as a result of compression, which favors evaporation. Water also evaporates from the cut surface on the outside of the chamber. Steps need to be taken to minimize these problems, specifically by loading the chamber rapidly, humidifying the chamber, and raising the pressure slowly. In addition, you should plan to take the chamber to the plant to be sampled and avoid the temptation to carry the plant or excised sample to the pressure chamber. Not only is loading more rapid, but the plant environment is unchanged and the water potential of the intact system is more readily preserved in the sample.

LEAVES

Most leaves equilibrate rapidly with the applied pressure and are favored material for pressure chamber measurements. Enclose as much of the leaf as possible inside the chamber because pressure chambers measure tensions that extend throughout the sample and pressure must similarly extend over the whole sample insofar as is practical. This ensures that any deformation caused by tensions in the intact plant will be reproduced by the pressure in the chamber.

Leaves having petioles require a round seal but grasses and certain conifer needles require a slit seal. Sampling is similar for both except, for wide grass leaves having a large midrib, the blade is sampled on one side of the midrib. In this case, use a sharp razor blade to cut toward the midrib perpendicularly (Fig. 2.7A). Grip the tissue on the apical side of the cut and tear toward the leaf tip (Fig. 2.7B). This

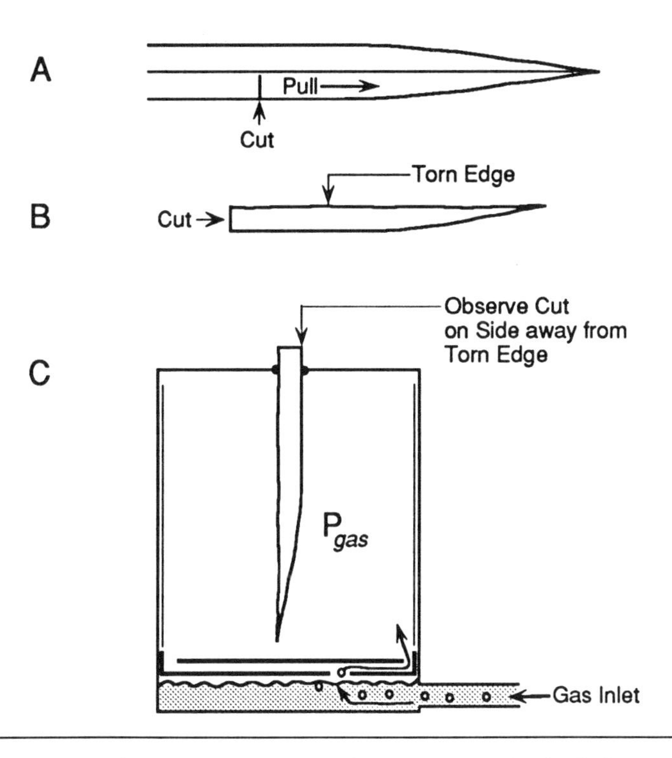

Figure 2.7. Sampling a wide grass leaf. Cut across the leaf almost to the midvein (A), tear the leaf toward the apex from the cut (B), and insert the sample in the pressure chamber (C). Observe the appearance of the xylem solution at the cut surface during pressurization but ignore solution appearing close to the torn edge.

gives a triangular sample with a damaged edge paralleling the midrib. Observe the xylem solution at the cut end but ignore the early appearance in the veins closest to the damaged edge (Fig. 2.7C).

Because leaf tissue is soft and easily crushed, avoid pressure chamber designs with pressure-activated seals where there is no control over how much force the seal applies to the tissue. A slight amount of crushing generally occurs and does not affect the balancing pressure but prolongs the time necessary to make the measurement. If crushing appears severe, test whether the xylem has been affected by leaving the tissue in the seal and excising the edge of the leaf to expose the ends of the veins. Apply pressure to the leaf in a water-filled pressure chamber. The flow through the tissue in the seal should be much faster than that observed with the intact sample, indicating that the xylem was not constricted by the seal.

Crushing the tissue also can release solution from the crushed cells. This adds liquid to the xylem solution, which appears at the cut surface earlier than if the tissue had not been crushed. Test this effect by loading a leaf into the seal in the usual way, but excise the blade before pressurizing the sample. Pressurize and tighten the seal. Any released liquid will appear on the cut surface and must be from crushing by the seal because there is no other tissue in the chamber. Crushing must be considered whenever the flow of water in or out of the leaf is important especially in pressure activated seals. Sometimes pressure chambers are used to study the rate of water release from leaves (Boyer, 1974; Koide, 1985; Tyree *et al.*, 1981) and care should be taken to avoid pressure activated seals in such studies.

BRANCHES

For woody branches, it may be desirable to strip away the tissues outside of the xylem for a short length so that only woody tissue is inserted into the seal. Regardless of the length of the branch, enclose as much as possible in the pressure chamber.

Branches contain a significant amount of nonleaf tissues such as pith, cortex, cambium, and so on. These often equilibrate slowly with the vascular tissue and pressure readings may be too rapid for complete equilibration. The effect can be demonstrated by measuring the water potential of the intact branch, then removing each leaf, placing Vaseline over the cut surfaces to retard gas entry, and repeating the measurement. If the branch without leaves gives a water potential different from that of the intact branch, equilibration of the entire branch did not occur during the intact measurement. One must then choose whether a "leaf balancing pressure" or "branch balancing pressure" is desired. The leaf balancing pressure is usually achieved in minutes but the whole sample balancing pressure may require hours or days.

ROOTS

Roots are often too fragile to measure individually unless there is extensive secondary thickening. Therefore, one usually uses a whole root system after detaching the shoot and seals the stump of the detopped stem in the chamber. Remove the root medium by placing the root system in a water vapor-saturated glove box and gently allowing the medium to fall away. Assemble the chamber without exposing the roots to dry air. Be sure that the chamber has been prehumidified and note the balancing pressure in the same fashion as with leaves or branches.

ROOTS IN SOILS

It is possible to obtain an average water potential for a root-soil complex by leaving the soil attached to the root system. Pressurize the sample in the same way as with other tissues.

Significant gradients in water potential can be present in the root-soil complex, particularly next to the root surface. The movement of water during pressurization collapses these gradients by forcing water from the bulk soil to the root surface. Thus, pressure readings tend to be weighted toward the potential of the bulk soil. Also, pressurizing wet soils can force water into the intercellular spaces of the root tissues with unpredictable effects (Passioura, 1984). There may be salt gradients next to roots when transpiration is rapid (Kramer and Boyer, 1995) and these can affect the pressure chamber readings.

As with leaves and branches, the pressure chamber measures the tension arising from the matric potential in the apoplast (Eq. 2.4). However, an endodermis separates the stele from the cortex and has Casparian strips that create a hydraulic barrier, and water probably flows mostly through the protoplasm at this barrier. Thus, the apoplast tension measured with the pressure chamber may extend only into the stele, and the cortical apoplast may be under much less tension.

Because soil contains solutes that affect the water potential of the roots, the selectivity of the root system for water is important. Assuming completely selective roots, the water potential of the root is obtained by determining the osmotic potential of the solution in the root xylem and adding the matric potential of the stele of the roots (Eq. 2.4). If the Casparian strips are not completely selective, however, a correction may need to be applied (see Significance of Reflection Coefficients, Chap. 4).

Roots in hydrated soils often will exude liquid onto the cut surface without any pressure application. This is a normal expression of root pressure, and pressure chamber measurements cannot be made. In dry soil, no exudation takes place and pressure measurements become possible.

Measuring the Components of the Water Potential

With the pressure chamber, the components of the water potential can be measured in the tissues surrounding the xylem provided it is ensured that there is equilibrium between the xylem and the rest of the tissues. The method allows the tissues to remain completely intact in the excised sample, which is an advantage.

OSMOTIC POTENTIAL

When solute is added to water, the free energy of the water decreases because the solute occupies space otherwise occupied by water, diluting the water and decreasing its chemical potential. As discussed in Kramer and Boyer (1995), Ψ_s approximates $-RTC_s = -RTn/V$ for dilute ideal solutions of nondissociating solutes. The C_s is the molar concentration of solute given as n/V ($mol{\cdot}m^{-3}$ of water), R is the gas constant ($m^3{\cdot}MPa{\cdot}mol^{-1}{\cdot}K^{-1}$), and T is the temperature (K). This relationship shows that Ψ_s is proportional to the solute concentration, and the pressure chamber can be used to remove water from the cells, leaving the solutes behind and making the cell solution more concentrated. As long as temperature is constant and the number of moles of solute n is a constant inside the cells, $-RTn$ is a constant (k):

$$V{\cdot}\Psi_s = -RTn = k, \qquad (2.10)$$

Rearranging Eq. 2.10 gives the equation of a line with a slope of ($1/k$):

$$\frac{1}{\Psi_s} = \frac{1}{k}{\cdot}V. \qquad (2.11)$$

The pressure chamber removes water from the cells by overpressuring them and, as the water moves out, the solution in the walls is diluted and its osmotic potential approaches zero. Equation 2.8 becomes $-P_{gas} = \Psi_{w(p)}$ and, with water loss from the cells, turgor becomes zero so that $-P_{gas} = \Psi_{w(p)} = \Psi_{s(p)}$. Replacing Ψ_s in Eq. 2.11 with $-P_{gas}$ gives

$$-\frac{1}{P_{gas}} = \frac{1}{k} \cdot V. \qquad (2.12)$$

Thus, in turgorless tissue, the osmotic potential of the cells is directly measured with the pressure chamber, and a plot of $-1/P_{gas}$ versus V gives a straight line of slope $(1/k)$ because of the concentration dependence of the osmotic potential (Fig. 2.8).

PROCEDURE

1) Apply an overpressure to the tissue, drive out a small volume of water, collect the volume in a syringe, and note the volume.
2) Adjust the pressure to the new balance, and note the pressure. This gives the pressure at the new water volume in the cells after removing the water in 1).
3) Repeat 1) and 2) for about 10 water contents.
4) Determine the final water content of the tissue by releasing the pressure, removing the tissue, excising the veins and any stem, weighing the interveinal tissue, and oven drying the tissue.
5) The difference between the weight of the interveinal tissue before and after oven drying is the volume of water in the tissue at the end of the pressure series (the veins and stem are usually removed because their water content is considered to be relatively constant).
6) Add each removed volume to the volume in the tissue at the end of the pressure series and express each sum as the relative water content at each balancing pressure (Richter, 1978). The relative water content is relative to the maximum water content in fully hydrated tissue expressed as a percentage (see Fig. 2.8). The relative water content is a measure of V in Eq. 2.12.
7) Plot $-1/P_{gas}$ versus the relative water content (Fig. 2.8). The initial part (low pressures) is not straight, but the final part (high pressures)

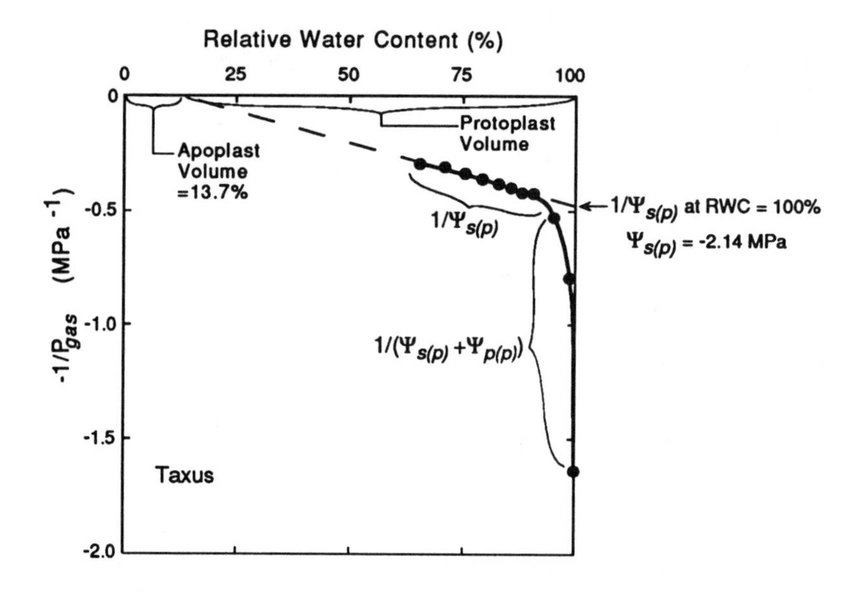

Figure 2.8. Pressure·volume determination for a *Taxus* branch in a pressure chamber. The relative water content is the volume of water in the tissue relative to that in a fully hydrated sample. Overpressures remove water from the leaf in steps, and the balancing pressure and removed volume are noted at each step (individual data points). The relative water content is determined by measuring the volume remaining after all overpressures are finished and adding the volumes that were removed by each overpressure. The dashed line shows the linear relationship governed by the osmotic potential ($\Psi_{s(p)}$, Eqs. 2.11 and 2.12). Extrapolation to the axis on the right gives the osmotic potential at 100% relative water content, and to the axis above gives the apoplast volume. The curvilinear part of the pressure·volume relation (on the right at high relative water contents) shows the effect of the osmotic potential plus the turgor pressure ($\Psi_{s(p)} + \Psi_{p(p)}$) inside the cells. The $\Psi_{s(p)}$ determined from the dashed line is subtracted from the $\Psi_{s(p)} + \Psi_{p(p)}$ to give the turgor (see Fig. 2.9). Data from J. S. Boyer (unpublished).

becomes straight. The straight portion represents the region where Eqs. 2.11 and 2.12 are followed and $-1/P_{gas} = 1/\Psi_{s(p)}$ or, in other words, where $-P_{gas}$ directly measures the osmotic potential of the tissue. Extrapolating the straight portion of the line to other water contents indicates the osmotic potential at any other water content (Fig. 2.8, dashed line). Extrapolating the straight line to the X axis shows the volume of water remaining after all the protoplast water has been removed (P_{gas} becomes infinite, Fig. 2.8). The remaining water is the volume of the wall and xylem water (apoplast volume), considered to be incompressible.

ASSUMPTIONS

Other ways of analyzing P·V data have been suggested but these are generally less satisfactory than the approach in Eq. 2.12 (Tyree and Richter, 1982). All methods rest on the assumption of a constant solute content in the cells, and Kikuta and Richter (1988) point out that the content may not be constant if solutes are generated by the cells during pressurization. This is not a problem with most tissues but wheat leaves appear capable of enough solute generation to cause an error (Kikuta and Richter, 1988). In general, the cell solutions are assumed to be so dilute that concentrations can be used instead of activities to express solution properties (Tyree and Richter, 1981).

Caution needs to be used in the extrapolation of the straight line. The extrapolations in Fig. 2.8 assume that all the water is released from the cells and none from the walls and xylem. The wall pores and xylem usually are reasonably rigid except in young tissue (Tyree, 1976), but significant water released by these structures can cause extrapolation errors. In particular, the extrapolation to the x axis is long and thus there is a considerable degree of uncertainty (see Changes in Xylem Dimensions).

The method also assumes that the cell walls around the xylem are not rigid and will collapse into the cell compartment without resistance as the tissue dehydrates. In tissues with stiff walls, the collapse may be resisted and $-P_{gas}$ may not equal $\Psi_{s(p)}$, necessitating a correction (see Negative Pressures Inside Protoplasts).

You can test whether any of these problems affect your data by calculating $\Psi_{s(p)} \cdot V$ at various V, that is, $\Psi_{s(p)}$ multiplied by the relative water content at any relative water content. This is a measure of the solute content of the tissue (Eq. 2.10) and establishes whether the solute content has remained constant over a wide range of water contents, as required by the theory. It also shows whether the xylem and cells follow solution behavior. The $\Psi_{s(p)} \cdot V$ should not vary along the straight line portion of Fig. 2.8. It is wise to make this test on all measurements of Ψ_s.

Another feature of the method requiring caution is the considerable time needed for making a series of overpressures. It is important to use compressed air so that oxygen is available to the tissue during the measurements and to minimize evaporation by humidifying the chamber, collecting the removed liquid quickly, and determining each new balancing pressure quickly (but be sure to wait long enough for a true balance). Avoid pressures above 3 to 4 MPa if possible because cell membranes can be disrupted and release cell solutes to the vascular system, which will cause the plot to deviate from linearity. If you have difficulty achieving linearity, measure the osmotic potential of the solution exuding from the cut surface. An osmotic potential significantly below zero at high pressures means that cell membranes have broken and cell solutes are being released to the apoplast. In this situation, the measurement must be abandoned.

TURGOR

Turgor results in pressure on the cell solution, and the balancing pressures in a pressure chamber are less than in comparable turgorless tissue. When turgor is present, the balancing pressures do not follow Eqs. 2.11 and 2.12, and a plot of Eq. 2.12 curves downward (on the right close to the Y axis, Fig. 2.8). By comparing the downwardly deviating line (which describes $-1/P_{gas} = 1/(\Psi_{s(p)} + \Psi_{p(p)})$ with the linear extrapolation (which describes $-1/P_{gas} = 1/\Psi_{s(p)}$), the turgor can be determined by difference (Eq. 2.3), and the turgor can be found at any tissue water content (Fig. 2.9).

MATRIC POTENTIAL

Matric potentials occur because the surface of a liquid has properties that differ from those in the interior. In any porous medium wettable by water, solids extend the surface so that a larger share of the molecules have surface properties. The wettability results mostly from hydrogen bonding between water and OH groups on the surfaces and from surface charges that attract the water dipole. The surface charges also attract ions in the water. The total effect is to constrain water and solute next to the surfaces. The wettability attracts water from the air and any liquid water and, because water forms strong bonds with other water molecules, the pores in the matrix tend to fill. Electrically constrained ions next to the pore surfaces also move water into the matrix with osmotic-like force. As a consequence, the water content of the matrix can become very large. Pressures are generated next to the surfaces and the whole matrix can swell.

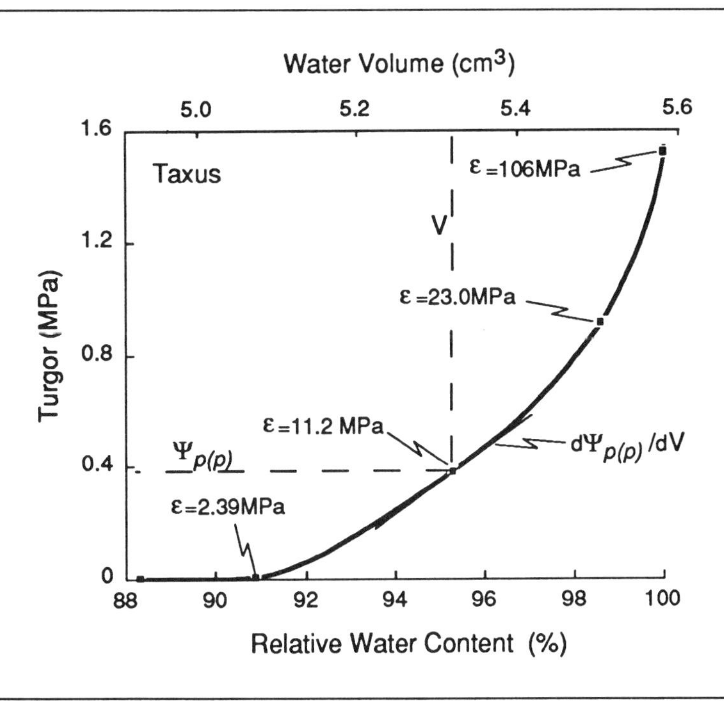

Figure 2.9. The turgor ($\Psi_{p(p)}$) and volume (V) of water in the cells of the *Taxus* branch shown in Fig. 2.8. The slope $d\Psi_{p(p)}/dV$ was measured at each $\Psi_{p(p)}$ and V, and the elastic modulus (ε) also was calculated. In the example shown, the slope is drawn through a point and ε = (0.380 MPa/0.180 cm^3)·5.32 cm^3 = 11.2 MPa. Data from J. S. Boyer (unpublished).

For most plant cells, the walls are the major site of the matric potential (Boyer, 1967b). The surfaces are highly wettable, and water fills the pores. Because of the small pore diameter, tensions (negative pressures) to about -58 MPa can be present without draining the water. Of course the tension on pore water varies between zero and -58 MPa, and the various pressures measured with a pressure chamber demonstrate this principle. Accordingly, pressure chambers give a direct measure of the matric potential in the walls of the living tissue ($\Psi_{m(a)}$ in Eqs. 2.4 and 2.7).

The pressure chamber also can measure the matric potential in leaves killed by freezing and thawing (Boyer, 1967b), which is sometimes useful (see Figs. 2.10 and 2.11). In this situation, there is no turgor and the osmotic potential is virtually without effect because there are no membranes. The only force holding water in the tissue is the matric potential resulting from surface interactions. As pressure is applied to the system, water and solute move out of the cells and exude from the cut end of the petiole. The walls tend to collapse into the cell compartment and resist collapse according to the strength of the wall, but the measured matric potential is accurate regardless of how much the walls collapse.

The forces holding water in the dead matrix are of the same physical nature as those holding water in the apoplast of living tissue. However, freezing and thawing flood the walls with protoplast solution and change the matric potential of the apoplast to a higher value (less negative) than in the living tissue (Boyer, 1967b). Only when the water content in the frozen/thawed tissue is the same as in the walls of living tissue do the matric potentials approach those in the living tissue, as pointed out by Boyer (1967b) and Passioura (1980).

With frozen/thawed tissue, it is usually desirable to keep the petiole alive so that it retains its usual strength in the seal in the top of the chamber. Wrap the petiole in wet cotton while the leaf blade is being frozen and, with care, a freezing time can be found that allows the blade to be frozen but not the petiole. Mount the petiole in the seal of the pressure chamber in the usual way and pressurize the tissue slightly. Increasing pressures will cause the cell solution to exude onto the cut surface and decreasing pressures will cause the exudate to move back into the tissue. Because the matrix is flooded with cell solution released by freeze/thawing, the pressures are small and rates of exudation are slow.

ELASTIC MODULUS OF PLANT TISSUE

As turgor is generated by the walls pressing on the cell contents, the walls are under strain much as the cover of a ball comes under strain at high internal pressures. The strain is elastic and reversible. The elasticity of the strained cell wall can be measured at various water contents using a pressure chamber (Fig. 2.9). Because pressure applies a force in three dimensions, the elasticity is determined as the bulk modulus of elasticity (ε, MPa) defined by

$$\varepsilon = \frac{d\Psi_{p(p)}}{dV} \cdot V \tag{2.13}$$

or

$$d\Psi_{p(p)} = \varepsilon \cdot \frac{dV}{V}. \tag{2.14}$$

The bulk modulus is a proportionality constant indicating how much change occurs in the relative cell volume dV/V when the pressure inside the cell changes by an amount $d\Psi_{p(p)}$. The more elastic the cell wall, the smaller is the value of ε. The ε is maximum when turgor is at its maximum but becomes zero when the turgor is zero for a cell having thin elastic walls. Figure 2.9 shows this effect using data from Fig. 2.8. Because the pressure chamber measures volume-averaged tissue $\Psi_{p(p)}$, the ε is a volume-averaged parameter.

Precautions

SAFETY

Pressure chambers can be dangerous because of the large gas volumes involved. There can be explosive release of the tissue from the seal or failure of a chamber component. Always work at the side of the chamber and never view the cut surface from overhead (Fig. 2.6). To avoid the failure of a chamber component, pressure test the sealed and water-filled chamber before the first use as described earlier. All components should withstand pressures at least double the maximum expected to be used.

DEHYDRATION DURING LOADING

Excised tissue continually dehydrates. In normal air, its potential decreases significantly in a few seconds. It is generally impossible to prevent this completely but placing the tissue in a saturated atmosphere and keeping it as isothermal as possible minimizes the problem (plant tissue produces small amounts of heat metabolically and thus loses water slowly even in saturated air).

With these principles in mind, one must work swiftly and place the pressure chamber next to the plant to minimize the time between excision and chamber loading. Because the chamber contains saturated air, loading reduces the rate of evaporation. If loading times are longer than 10 sec, quickly place the tissue into a glove box with a saturated atmosphere so that loading can be extended without dehydrating the tissue. A convenient glove box can be constructed from a Styrofoam box with a Plexiglas sheet on top (see Chap. 3). An alternative is to quickly place a thin polyethylene bag around the sample immediately prior to excision (Turner and Long, 1980). Seal the bag loosely, shade, excise the tissue, and insert the sample into the seal of the chamber top. Apply pressure to the tissue in the bag.

DEHYDRATION DURING PRESSURIZATION

After sealing in the chamber, the tissue is exposed to dehydration from the dry gas used to pressurize the sample. Pressure chambers should be humidified by bubbling the incoming gas through water in the bottom and past walls lined with wet filter paper. The water in the bottom not only humidifies but also cools the incoming gas.

A baffle can be inserted in the pressure chamber just above the liquid to prevent water droplets from splashing onto the tissue (Fig. 2.5). A simple baffle can be made from the bottom of a suitably sized polyethylene bottle which is forced into the pressure chamber and held by the walls just above the surface of the water. Place one or two small holes in this false bottom to allow compressed air to move into the main body of the chamber. From another plastic bottle, cut a flat disc that is slightly smaller than the inside diameter of the chamber. Place this disc on top of the false bottom. The incoming air will bubble through the water, pass through the small holes in the false bottom, and go around the disc to enter the main portion of the chamber. The disc breaks up any water droplets that are blown through the holes. If your pressure chamber is not designed for gas to enter the bottom, insert a tube inside the chamber to lead the gas from the inlet to the bottom.

MEASURING AT EQUILIBRIUM

Pressure chambers generally use large tissue samples. Because there are water potential gradients in plants and soils (see Boyer *et al.*, 1980, for an example), it is important to make measurements slowly enough to allow the gradients in the sample to equilibrate and form a volume-averaged potential. For the most part, gradients in leaf samples will equilibrate adequately in 10 min. In larger samples (e.g., branches), the time is longer.

Pressure chambers are useful for making water potential measurements rapidly, and it is tempting not to wait for a true balancing pressure. Indeed, some pressure chambers are designed to allow air to enter at a steady rate and be shut off at the first sign of liquid on the cut surface. This method is not recommended because it does not allow balancing pressures to be achieved. Moreover, the tissue can be heated by rapid air entry into the chamber (Puritch and Turner, 1973). To ensure that equilibrium occurs, the pressure should be adjusted to give a stable, flat water film at the cut xylem surface. The film should not grow or shrink, indicating that water is neither exiting nor entering the tissue.

OVERHEATING

When pressure chambers are used continuously at high ambient temperatures, spurious readings can result. The problem is most often encountered at temperatures above 30°C in the field when large amounts of heat are produced by frequent compression of the incoming gas (Puritch and Turner, 1973). The errors in the readings are caused initially by excessive dehydration of the tissue and ultimately by breakdown of the membranes because of the heat. The breakdown causes readings to be too low. The effect can be minimized by wrapping a wet paper towel around the outside of the chamber to allow evaporative cooling of the chamber walls. Also, pressurizing slowly and bubbling the gas through water on the chamber bottom will help keep the chamber cool.

AVOIDING TISSUE HYDRATION BY SURFACE WATER

Plant tissue sometimes may be coated with dew or have droplets of water on the surface. When pressurized, the liquid water is forced into the tissue and raises its water potential. If water is present on the tissue to be pressurized, blot the tissue dry. If possible, allow some time for the tissue to dry completely before sampling. Avoid tissue contact with wet filter paper on the walls by folding the leaf loosely and holding the folds in place with a rubber band or tape. Water splashing from the bottom of the chamber can be avoided by constructing a baffle in the bottom (Fig. 2.5).

ANATOMICAL ERROR

Pressure chambers use the cut surface of the sample as a reference position for the measurement. Therefore, never recut the tissue after removing the sample from the plant. Tissues comprising the cut surface can have hollows or pith that allow the xylem solution to spill over and be trapped after it first appears on the cut surface (Boyer, 1967a). More pressure is necessary to maintain the liquid at the surface in this situation and the reading will be spuriously high. In this case, pressure readings are considered to be relative rather than absolute measurements (Boyer, 1967a).

BUBBLING IN XYLEM SOLUTION

In a typical sample, pressurization does not cause much movement of gas through the tissue. Of that appearing at the cut surface, most travels through the intercellular spaces. A small amount moves through the xylem probably because gas has been forced into solution at high pressure and bubbles out of solution as atmospheric pressure is encountered.

Wounding of pressurized samples can allow gas to enter the vascular tissue more rapidly. Liquid in the xylem appears prematurely at the cut surface and bubbling can be so severe that the water films are difficult to observe or are dissipated as a fine spray. Wherever possible, avoid using wounded tissue for measurements. If wounds cannot be avoided, it is sometimes possible to coat the wounded area with petrolatum (Vaseline) and decrease the rate of gas entry.

Xylem solutions sometimes contain compounds that have surfactant properties, and the solution foams on the cut surface. If foaming is excessive, the measurement must be abandoned. If it is moderate, increase the pressure until liquid accumulates underneath the foam on the cut surface. Then withdraw the liquid into the tissue and determine the balance pressure. If the foam fails to break up during this procedure, touch it with your finger.

CHANGES IN XYLEM DIMENSIONS

Mature plant xylem in secondary tissues withstands large pressures. Immature xylem, protoxylem, and metaxylem do not nor do tissues around the xylem, and their volume decreases as water flows out of the cells. The xylem is often under large tensions, and when the tissue is excised and pressurized, some of the xylem can change dimensions or can exchange water with the surrounding tissues, particularly if the sample contains growing tissues. To avoid error caused by these effects, the xylem should be pressurized along as much of its length as possible to force the surrounding tissue to the same water content during the measurement as it had when the xylem was under tension in the plant. You can test whether these effects are a problem by changing the position of the sample in the seal. Use alternating measurements with most of the xylem pressurized (most of stem inside of chamber) or the least amount of xylem pressurized (most of stem outside). If balancing pressures are less when most of the xylem is pressurized, stem tissues are sensitive to pressure and need to be considered.

POTENTIAL GRADIENTS

In addition to the need to let potential gradients equilibrate in the sample, there are long-distance gradients that need to be considered (e.g., Boyer *et al.*, 1980). The two largest contributors to gradients are the distance of the sample from the water supply and the degree of illumination of the sample. In most cases, the variability between samples can be markedly reduced by knowing where gradients exist and by sampling in the same part of the gradient. Comparisons of potentials between plants under these conditions require samples from the same part of the canopy, similarly illuminated, and at a similar stage of development.

Depending on the research question, one should sample at an appropriate position in the plant gradient. For example, to measure the water potential of leaves that are doing most of the photosynthetic work for the plant, sample at the top of the canopy using leaves that are recently fully expanded and oriented perpendicular to the incoming light.

Enclosing leaves in gas exchange cuvettes poses special problems because the enclosure generally changes the leaf water potential. Always measure the water potential of the leaf in the cuvette if you wish to relate leaf performance in the cuvette to the water status of the leaf. Avoid sampling leaves outside the cuvette and assuming that the water potential is the same as inside the cuvette. If you cannot sacrifice the leaf in the cuvette for the pressure chamber measurement, consider using a thermocouple psychrometer on a small leaf sample (see Chap. 3).

NEGATIVE PRESSURES INSIDE PROTOPLASTS

Pressures rise and fall in cells according to changes in cell water contents. Turgor is high in the protoplasts when water contents are high, and tensions are small in the apoplast. As water contents decrease, turgor diminishes and tensions become greater. A question then arises: if water contents continue to decrease, does turgor disappear and tension begin to extend into the protoplasts and, if so, does this affect pressure chamber measurements? The answer appears to be yes for both questions under particular conditions.

The effect depends on how much the cell walls resist collapse under tension. In most tissues, the cell walls are thin and follow the shrinkage of the cell solution without significant resistance as water contents decrease. In sunflower, for example, the walls occupy only 9-12% of the total cell volume (Boyer, 1967b), and they tend to fold and follow the shrinking protoplasts as the cells dehydrate, which is clearly visible under the electron microscope (Fellows and Boyer, 1978). This suggests that they do not exert a significant counterforce to the shrinkage, and Fig. 2.10 shows that the volume changes readily in frozen/thawed tissue, and the matric potential ($\Psi_{m(t)}$) is small when compared to that in the living tissue ($\Psi_{m(a)}$). The frozen/thawed tissue has no turgor and the small $\Psi_{m(t)}$ indicates that the walls do not resist shrinkage as water is lost from the matrix. This lack of much counterforce continues until relative water contents fall below 10-20%. Accordingly, at a higher water content of 60%, the free energy diagram on the right of Fig. 2.10 shows that $\Psi_{m(t)}$ is a very small component compared to P_{gas} in the living tissue. Therefore, in living sunflower, at these higher water contents, tensions in protoplasts can be neglected for pressure chamber measurements.

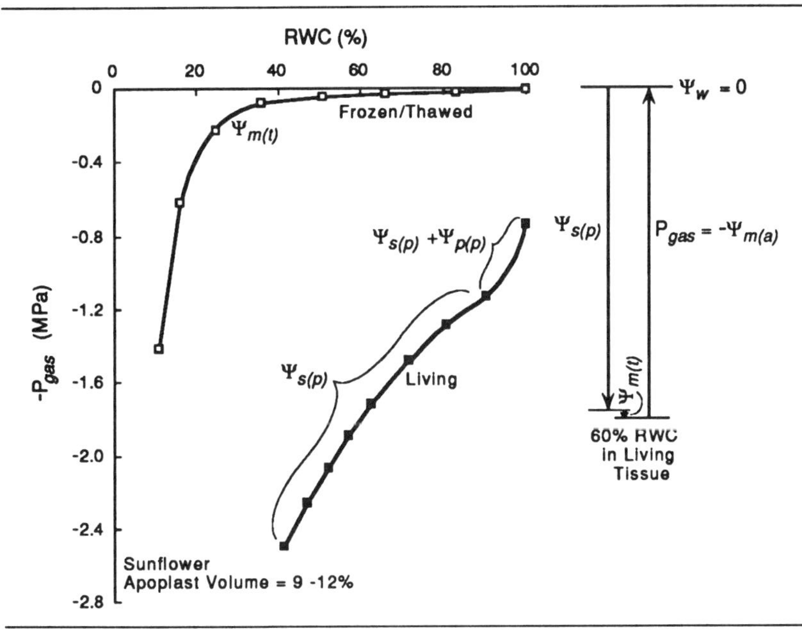

Figure 2.10. Comparison between pressure-volume data in living and frozen/thawed leaves of sunflower. In living tissue, the pressure chamber ($-P_{gas}$) measures the matric potential $\Psi_{m(a)}$ of the apoplast (Eq. 2.7). In frozen/thawed tissue, the pressure chamber measures the matric potential $\Psi_{m(t)}$ of the entire nonliving tissue. The $\Psi_{m(t)}$ is mostly caused by the cell walls. Note that between 40 and 100% RWC, $\Psi_{m(t)}$ changes little and large amounts of solution can be removed indicating that the walls collapse into the shrinking cell compartments. The cell walls occupy 9-12% of the total cell volume and thus have little resistance to the shrinkage. The diagram on the right gives the components measured by the pressure chamber in living tissue at 60% RWC taken from the graph on the left. In living tissue, a 60% water content is unable to generate turgor, and the $-P_{gas} = \Psi_{m(a)} = \Psi_{s(p)}$. The $\Psi_{m(t)}$ is so small that it can be neglected. Data from J. S. Boyer (unpublished).

However, tissues having thick cell walls behave differently. Figure 2.11 shows that in rhododendron, the apoplast volume was 26 to 28% of the total water volume in the tissue (Boyer, 1967b). The walls were relatively rigid and resisted collapse and in consequence the frozen/thawed tissue showed a substantial $\Psi_{m(t)}$ at all water contents. Therefore, when turgor disappeared in living leaves, the tension in the xylem extended into the surrounding protoplasts.

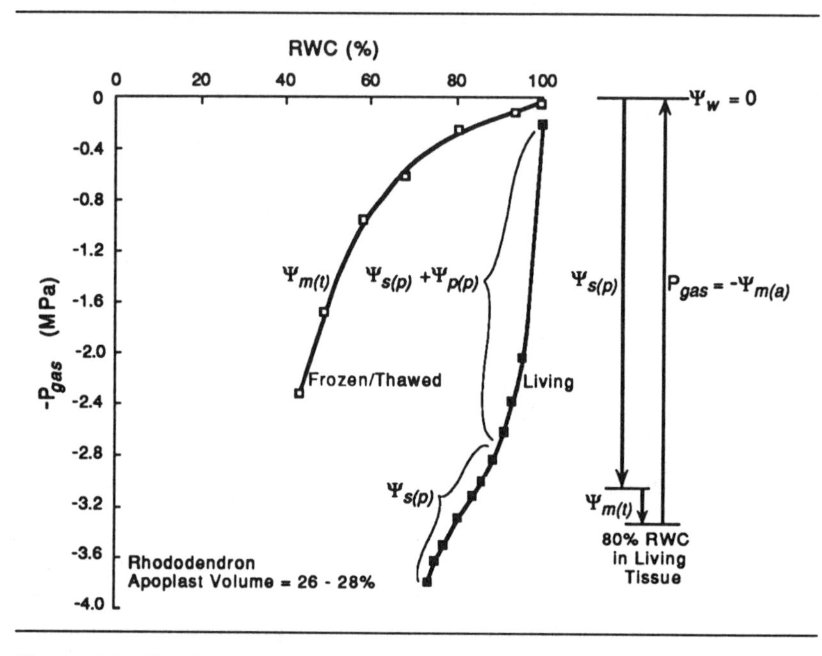

Figure 2.11. Same comparison as in Fig. 2.10 but for rhododendron instead of sunflower. Rhododendron leaves have stiffer cell walls (26-28% of cell volume) than sunflower (9-12% of cell volume), and $\Psi_{m(t)}$ is a significant component of $-P_{gas}$ at most water contents (e.g., diagram on right for 80% RWC). As shown on the right, $\Psi_{m(t)}$ contributes so much to the measurement of $-P_{gas}$ that $\Psi_{m(t)}$ must be subtracted from $-P_{gas}$ to obtain a valid measure of $\Psi_{s(p)}$ in living tissue. This becomes important for measurements of $\Psi_{s(p)}$ in any tissue having rigid cell walls. For a test to determine whether the subtraction is necessary, see the text. Data from J. S. Boyer (unpublished).

A simple test for this effect is to determine whether $\Psi_{s(p)} \cdot V$ is constant over the linear range of volumes in Eq. 2.12 (see Measuring the Components of the Water Potential, Osmotic Potential). If it is not, a possible cause is a tension in the protoplast compartment caused by the resistance of cell walls to collapse during dehydration of the tissue. The tension can be subtracted from the P_{gas} to give a more accurate osmotic potential (Fig.2.11). Freeze/thaw a comparable sample and, from measurements of $\Psi_{m(t)}$ at various relative water contents as shown in Fig. 2.11, subtract the pressure used to measure $\Psi_{m(t)}$ from P_{gas} in the living tissue (Fig. 2.11 on the right). This subtraction should give a corrected $\Psi_{s(p)} \cdot V$ that is constant and thus provide a true $\Psi_{s(p)}$ at each water content.

It should be recognized that in the living tissue, the tension begins to extend into the protoplasts only after turgor decreases to zero. Tyree (1976) recognized that this tension could be present but considered it to be a "negative turgor" (an unfortunate misnomer) that was negligiblc in most cases. His test was based on the straightness of the slope of the line generated by the data, as in Fig. 2.8. He recognized that a better test would have been $\Psi_{s(p)} \cdot V = k$ to indicate that the solution in the cells behaved ideally. However, the test could not be made because the required data were not available. For the measurements shown in Figure 2.8, however, the test could be made and $\Psi_{s(p)} \cdot V = k$ at all relative water contents showing that negative pressures were not a factor.

Appendix 2.1-Pressure Chamber Manufacturers

Skye Instruments, Inc.
P.O. Box 278
Perkasie, PA 18944
USA
Telephone: (215) 453-9484

PMS Instrument Company
2750 NW Royal Oaks Drive
Corvallis, OR 97330
USA
Telephone: (503) 752-7926

Pacific Agricultural Services, Inc.
4325 West Avenue
Fresno, CA 93722
USA
Telephone: (209) 275-0775

Soilmoisture Equipment Corporation
P.O. Box 30025
Santa Barbara, CA 93105
USA
Telephone: (805) 964-3525

Chapter 3

Thermocouple Psychrometry

Thermocouple psychrometry is the most widely used method of measuring plant water status and probably is the most versatile. It can measure the water status of any plant part as well as soil or any substance containing water. The technique works over the entire range of water contents and, because it measures conditions in the gas phase, it does not require a continuous liquid phase for the measurement. The only requirement is that water be able to evaporate from the sample to the air. The method uses only a small sample which can be important for repeated measurements in the same plant or soil.

At the same time, psychrometry requires a good deal of care in its use. The sample is sealed in a closed container and the water in the sample evaporates to humidify the atmosphere. The measured humidity must accurately reflect the condition in the sample. The vapor pressures usually are high and approach saturation (relative humidity of 100%). With the humidity so near saturation, temperatures must be uniform so that condensation does not occur and lower the humidity. For the best accuracy, temperatures should not differ by more than 0.001°C anywhere in the chamber. Also, thermocouples are used to measure the humidity, but at high humidities they produce only a few billionths of a volt. Therefore, not only must temperatures be uniform, but small voltages must be accurately detected.

Instruments that incorporate these principles are available commercially. They use insulation and specially constructed and shielded circuits to provide the required stability. Depending on the specific design, measurements can be made with an accuracy that can exceed ±0.01 MPa. In this chapter, we will assume you have a commercial instrument (Appendix 3.1) and will describe the procedures for using it. However, the principles apply to any psychrometer and should be useful with custom-made units. For additional information, see Boyer (1969b) and Brown and van Haveren (1972).

Principles of the Method

When any liquid is sealed into a chamber containing air, it will evaporate until the partial pressure of the vapor equals the vapor pressure of the liquid. For example, liquid water will evaporate, causing the humidity to rise (Fig. 3.1A). Eventually the humidity becomes so high that the vapor condenses into the liquid at a rate that

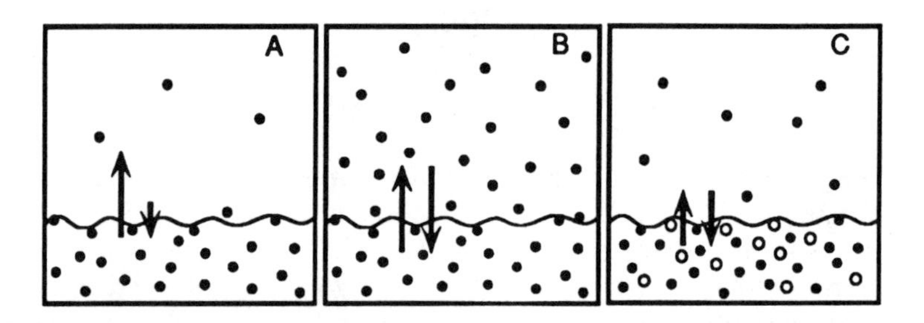

Figure 3.1. Evaporation of water in a closed container. A) Evaporation immediately after closing the container. The humidity is rising, and evaporation (upward arrow) is faster than condensation (downward arrow). B) At equilibrium, humidity has risen until evaporation occurs at the same rate as condensation. Humidity no longer changes. Relative humidity above pure water is 100%. C) For a solution, water molecules (solid circles) are excluded from the surface occupied by solute molecules (open circles). As a result, evaporation is slower and condensation slower (short arrows) and the humidity is lower than in B at equilibrium. The lower humidity (e_w /e_o) compared to B indicates a decrease in the free energy of the water molecules.

equals the evaporation (Fig. 3.1B). In this condition, the partial pressure of the vapor in the air equals the vapor pressure of water in the liquid and there is no further change in air humidity. The system in the chamber is in equilibrium because there is no net transfer of water within the system and the whole system remains stable with time. If we can measure the humidity of the air, we will know the vapor pressure of the liquid. As discussed in Chap. 1, the vapor pressure indicates the chemical potential of the water and thus the water potential.

Knowing the vapor pressure of the liquid tells us the water status of the liquid because the vapor pressure changes as various factors affect the molecules in the liquid. For example, adding solute to the water displaces some water from the space occupied by the solute (Fig. 3.1C). This will cause fewer water molecules to be exposed at the surface of the liquid and will decrease the rate of evaporation. A slower condensation rate will be required to equal the slower evaporation rate.

The simplest method of measuring the humidity is to place a solution of known vapor pressure into the atmosphere (Boyer and Knipling, 1965). If it evaporates, its vapor pressure is higher than the

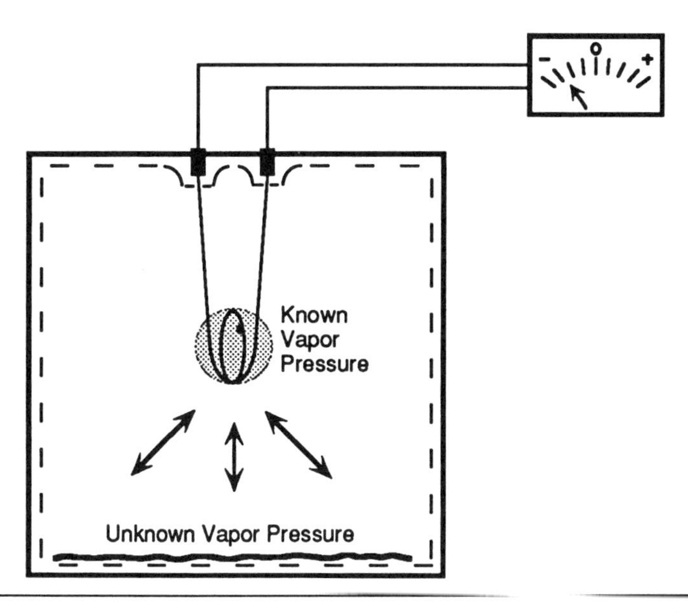

Figure 3.2. Evaporation from a thermocouple solution to a liquid of unknown vapor pressure inside plant tissue (wavy solid line on bottom). The thermocouple is wound in a spiral to form a loop where a droplet of solution can be placed (shaded area). The sensing junction is the black dot. If the thermocouple solution has a vapor pressure higher than that of the unknown, water evaporates from the solution and condenses in the unknown, cooling the solution, decreasing the voltage of the sensing junction, and decreasing the voltage on the meter, which remains steady as long as evaporation is steady.

humidity of the atmosphere. If it condenses water from the air, its vapor pressure is lower. Evaporation or condensation can be detected from the temperature. If evaporation occurs, the solution cools. If condensation occurs, the solution warms. The temperature can be detected with a thermocouple (Fig. 3.2). When the solution cools, the voltage produced by the thermocouple is low. When the solution warms, the voltage is high. When the voltage is the same as that of a dry thermocouple, which is neither cooled by evaporation nor warmed by condensation, the vapor pressure of the solution on the thermocouple is the same as the humidity of the atmosphere and in turn the vapor pressure of the solution in the chamber. The solutions are isopiestic, that is they have the same vapor pressure, and the isopiestic condition identifies the vapor pressure and thus the chemical potential of the solution in the chamber.

The isopiestic condition also is isothermal because the thermocouple solution and the sample solution have the same temperature, which is the same as the uniform temperature surrounding the chamber. As pointed out earlier, an isothermal condition is necessary before a relationship can be seen between the vapor pressure and the water potential.

Other variations of this method are available and will be described but all of them enclose a small sample in a vapor chamber and all use thermocouples to determine the humidity of the air above the sample. The main differences are in the voltage measurement for the thermocouple and whether calibration is required.

Theory of Psychrometry

As discussed in Chap. 1, there is a relationship between the vapor pressure of a solution and its water potential Ψ_w (Eq. 1.5). The relationship can be simplified to

$$\Psi_w = 137.2 \ln \frac{e_w}{e_o} \tag{3.1}$$

at 25°C which shows that vapor pressure is a sensitive indicator of the water potential. For example, the vapor pressure of pure water gives a relative humidity of 100% when it is sealed in a chamber kept at a uniform temperature (e_w / e_o is 1), and $\Psi_w = 0$. When the water is not pure, e_w / e_o is less than 1 and $\ln(e_w / e_o)$ is negative, and Ψ_w is negative. For a relative humidity of 99.3%, Ψ_w is -1.0 MPa. Clearly, humidities are high above most biological samples and soils even when the water potential is quite negative!

From the process of evaporation, it can be seen that water vapor moves toward regions of lower vapor pressure and thus toward more negative water potentials. In the psychrometer chamber, the vapor moves similarly and most solutes are nonvolatile, so the thermocouple solution acts as though it is separated from the tissue by a differentially permeable barrier -- the atmosphere in the vapor chamber -- that allows only water to move. The movement can be opposed by an equal vapor pressure in the same way that the movement of liquid water through a membrane can be opposed by an external pressure. The potential measured by the opposing vapor pressure expresses not only the work that the water can do but also the direction the water will move (Chap. 1). The movement of vapor toward more negative potentials is thus an expression of the same phenomenon in the liquid and is the key to water acquisition by plants.

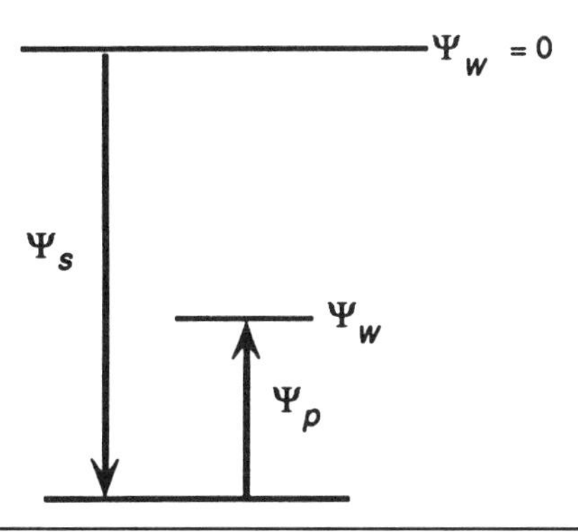

Figure 3.3. Free energy diagram showing the effect of solute (Ψ_s) and positive pressure (e.g., turgor pressure, Ψ_p) on the water potential of a solution inside a cell (Ψ_w). Solute decreases the free energy while positive pressure increases it. The top of the diagram shows the Ψ_w of pure, free water at the same temperature, which is the reference with a vapor pressure equivalent to 100% relative humidity and a water potential of zero.

Equation 3.1 is written in terms of the energy in the liquid-air system and not how fast the energy is achieved. The equal sign indicates that e_w / e_o will give an unambiguous measure of the energy when the measurement is made at equilibrium with an opposing vapor pressure. The equilibrium is achieved when the sensor (the liquid on the thermocouple) has the same energy, that is vapor pressure and temperature, as the liquid in the sample.

Vapor pressures change with temperature but in equilibrium measurements e_w and e_o are at the same temperature and the only effect is on the coefficient in Eq. 3.1. The coefficient 137.2 decreases to 131.6 when the temperature of the measurement decreases from 25°C (298 K) to 12°C (285 K), which is the same as the temperature dependence for the pressure chamber and pressure probe (see Chaps. 1, 2, and 4) and for solutions (see Appendix 3.2).

The vapor pressure develops at the surface of the solution in the sample. For plant tissue, this surface is in the cell walls and for soils it is on the surface of the soil particles. As discussed in Chap. 2, the water potential of the cell walls or apoplast is determined by the osmotic potential and matric potential of the apoplast solution, $\Psi_{w(a)} = \Psi_{s(a)} + \Psi_{m(a)}$, and a similar relation holds for the soil. The components are different for the protoplasts which have their own osmotic potential and a turgor pressure as shown in Fig. 3.3: $\Psi_{w(p)} = \Psi_{s(p)} + \Psi_{p(p)}$. Each protoplast is essentially in equilibrium with its cell wall: $\Psi_{w(a)} = \Psi_{w(p)}$.

The vapor pressure of the surface solution is affected by each of these components as water moves between the apoplast and protoplasts. Each component can be measured under the appropriate conditions in a psychrometer.

How Thermocouples Work

If two dissimilar metallic conductors are joined to make a circuit as in Fig. 3.4A, a voltage is generated at each junction (Seebeck thermoelectric effect). The voltage varies with temperature, but if the temperature is uniform, there is no current because the voltages in the circuit are the same and oppose each other (note that, in moving around the circuit, the sequence of metals at the first junction is opposite that at the other junction and thus the voltages are opposing). It is possible to insert a voltmeter to measure the voltage in the circuit (Fig. 3.4B). As long as the temperature remains uniform throughout the circuit, the thermoelectric properties of the voltmeter do not affect the circuit (the two contacts with the voltmeter generate equal and opposing voltages).

When the thermocouple junctions have different temperatures, the voltmeter will show the voltage difference between the two junctions with a sign that depends on which junction is warmer. In thermocouple psychrometers, one junction is held at the temperature of the surroundings, i.e., the temperature of the chamber in which the sample is located (T_1 in Fig. 3.4C), and is termed the reference junction (Fig. 3.5). The other junction is termed the measurement junction (Fig. 3.5) and is exposed to the chamber atmosphere where the junction is warmed (T_2 in Fig. 3.4C) or cooled (T_0 in Fig. 3.4D) by water condensing to or evaporating from the junction.

The reference junction usually consists of a pair of copper posts sealed into the top of the vapor chamber. The voltmeter is inserted between the two posts (Fig. 3.5). Because the posts have the same temperature (T_1), their effect on the voltages is equal but opposite and cancels. Thus, the circuit indicates the difference between the reference and measurement junctions as if the voltmeter was not there.

Thermocouples having their junctions at different temperatures absorb or liberate heat at the junctions in proportion to the current produced in the circuit (Peltier effect). The amount of heat absorbed or liberated can be increased by inserting a battery to increase the current flow around the circuit. The heat liberated at one junction equals the heat absorbed at the other junction, and reversal of the current flow reverses the liberation/absorption of heat at each junction. The Peltier

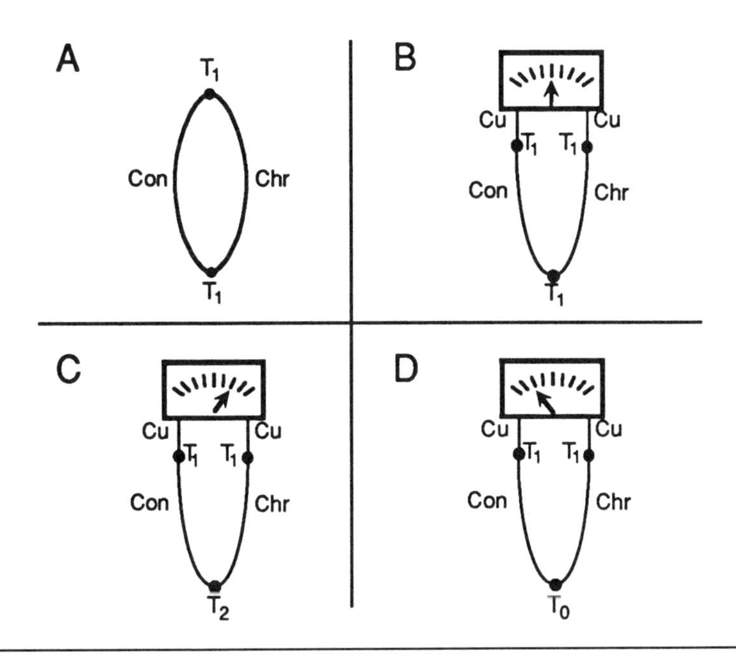

Figure 3.4. Thermocouple circuits. A) Basic thermocouple constructed of two dissimilar metals (Chr, chromel P; Con, constantan). Under isothermal conditions (T_1 at both junctions), the voltage is the same but opposing at the two junctions and no current flows. B) Same thermocouple as in A but with a voltmeter inserted at one junction. The voltmeter and wires are made of copper (Cu). In isothermal conditions (T_1), the voltmeter does not affect the circuit voltage and shows zero volts. C) Increasing the temperature of the lower junction (T_2) raises its voltage, and the voltmeter displays a positive voltage difference between the junctions at T_1 and T_2 (Seebeck thermoelectric effect). D) Decreasing the temperature of the lower junction (T_0) decreases its voltage, and the voltmeter displays a negative voltage difference between the junctions at T_1 and T_0.

effect is used to cool the measurement junction (and release heat at the reference junction) in some thermocouple psychrometers.

Types of Thermocouple Psychrometers

All psychrometers depend on the temperature of a thermocouple junction in contact with water or an aqueous solution

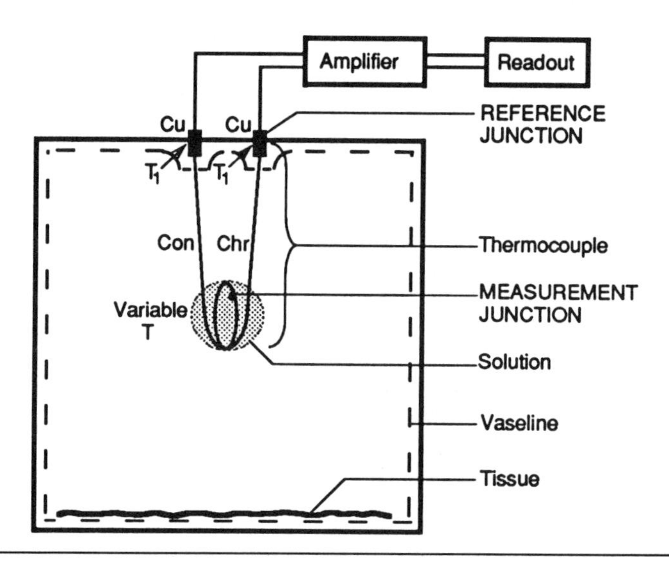

Figure 3.5. Typical thermocouple psychrometer circuit (Cu, copper; Con, constantan; Chr, chromel P). Reference junctions are at the copper posts, and the measurement junction is at the black dot. The voltage is measured between the posts at T_1 and the junction at variable T. Since both posts at T_1 have the same temperature, the voltage associated with copper cancels and only the voltage difference between T_1 and variable T is displayed by the amplifier and readout. Vaseline coats the chamber walls to minimize sorption of water vapor. The tissue sample is on the bottom.

suspended in the atmosphere above the sample (Fig. 3.2). According to the rate of evaporation, the thermocouple gives an electrical voltage. Because the output is small, it must be amplified (Fig. 3.5) before it can be sent to the voltmeter (readout device).

The amplifier introduces other junctions and switches into the circuit. These parts can act as thermocouples themselves and, to avoid interference with the measurements, they are kept isothermal. If there is doubt about a junction, a simple test is to warm it in your fingers and observe the effect in the readout of the instrument. If the junction is active there will be a change in the readout and extra steps should be taken to prevent temperature fluctuations around the junction.

It is essential that the walls of the psychrometer chamber are as nonsorptive as possible to facilitate vapor equilibrium. The most nonsorptive surface found so far is petrolatum (Vaseline) after it has been melted and resolidified to coat the chamber walls (Fig. 3.5). The melting and resolidification apparently hide impurities so that the surface becomes highly hydrophobic (Boyer, 1967a). Although some manufacturers do not include this coating in their procedures, it decreases interaction of the thermocouple with the chamber walls and has other beneficial uses (described later).

The detailed design of a psychrometer depends on whether it is used with samples removed from their surroundings, i.e., plant parts or soil samples, or *in situ*, e.g., attached to leaves or undisturbed parts of the soil profile. The sizes and shapes of the instruments can vary accordingly. However, the principles of the measurements fall into three categories: the isopiestic method, the dew point method, and the Peltier method.

ISOPIESTIC PSYCHROMETERS

Isopiestic means equal pressure and isopiestic psychrometers determine the vapor pressure of a known solution that equals the vapor pressure of the unknown (Boyer and Knipling, 1965). The thermocouple wire is bent to make a spiral loop that holds a droplet of solution (Fig. 3.6A) and contains the measurement junction. The water or solution is placed in the spiral, the thermocouple is inserted into the vapor chamber, and the output is observed. When the output is steady, the thermocouple is removed, the solution is cleaned off, a new solution is placed in the same spiral, and the thermocouple is reinserted into the chamber. The new steady output indicates the voltage change for the potential change caused by the new solution on the thermocouple. The potential of the solution can then be calculated that gives the output of a dry thermocouple, which is neither cooled nor warmed by evaporation or condensation. This solution is isopiestic and does not exchange water vapor with its surroundings. The water potential of the solution is known and is the same as in the sample, so no calibration is required. Diffusive characteristics of the sample do not affect the measurement, and the temperature is corrected simply by using the solution for the measurement temperature given in Appendix 3.2.

Any vapor pressure can be measured as long as a solution exists with a similar vapor pressure. The method has been tested for its accuracy (Boyer, 1966) by using plant tissue of known water potential and measuring the potential with the isopiestic technique (Fig. 3.7). The absolute accuracy was ±0.01 MPa; that is, the isopiestic value was within 0.01 MPa of the true water potential of the tissue and remained so for several hours. This is the only technique for which such a high accuracy has been demonstrated.

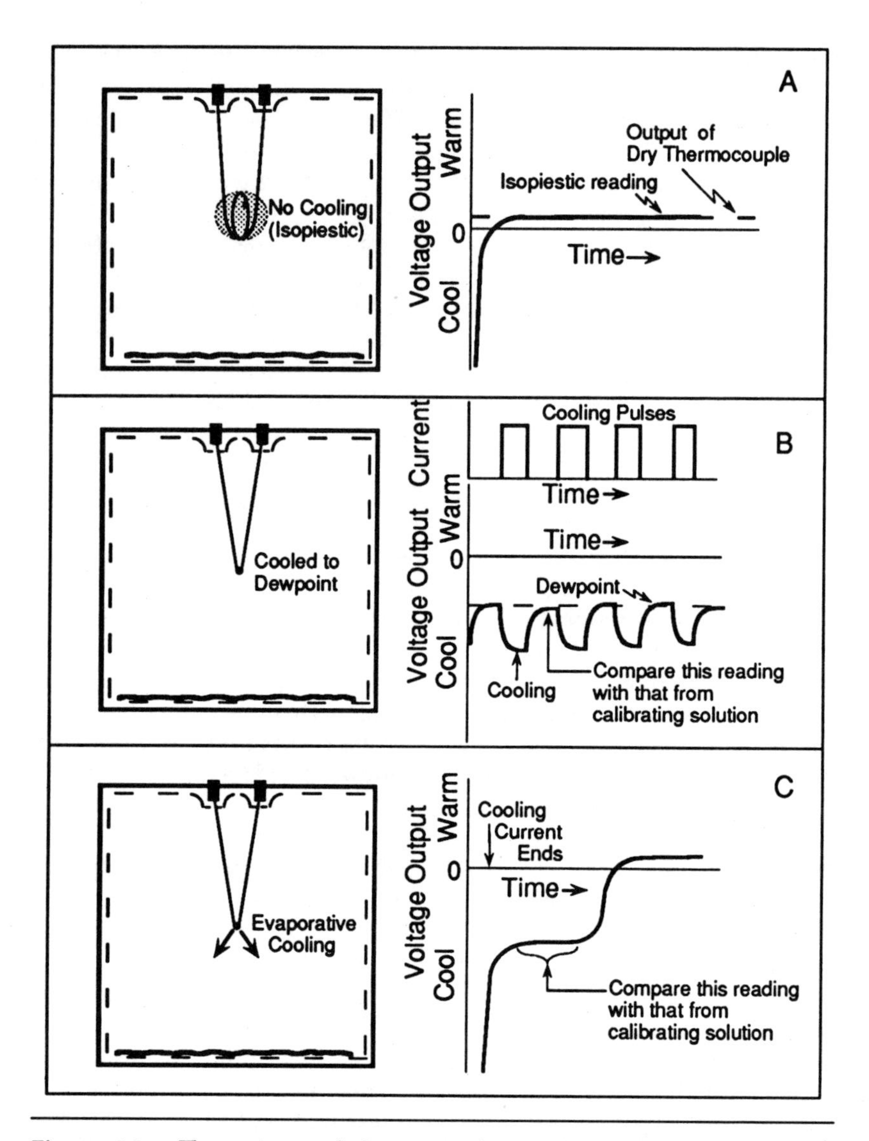

Figure 3.6. Three types of thermocouple psychrometers. A) Isopiestic psychrometer with a thermocouple that holds a droplet of solution of known vapor pressure (water potential). The thermocouple can be removed and the solution can be replaced with another having a different vapor pressure. The solution causing an output identical to that of the dry thermocouple is neither losing nor gaining water from the atmosphere and is thus isopiestic with the

DEW POINT HYGROMETERS

This method uses a Peltier current to cool the thermocouple just enough to keep the junction at the dew point (Neumann and Thurtell, 1972) where the condensed water has a vapor pressure that is the same as in the sample. The dew point is detected by determining the increase in Peltier cooling needed to cool the increased thermal mass of the measurement junction when it becomes wet. The dew point can be approximated by cooling in pulses and observing between pulses (Fig. 3.6B). This principle, described by Campbell *et al.* (1973), is used in a commercial unit (Appendix 3.1) that has an electrical circuit for automatically cooling and making the measurements (Fig. 3.6B).

Dew point instruments need to be calibrated with a range of solutions of known water potential (Appendix 3.1). The unknown dew point is compared with the dew points on the calibration curve to find the water potential of the unknown. The calibration is sensitive to ambient temperature.

Reproducibility is usually ±0.05 MPa (Campbell and Campbell, 1974; Nelsen *et al.*, 1978). With the commercial unit (Appendix 3.1), the true dew point is only approximated and there can be significant diffusion of water vapor between the sample and the junction, causing diffusive error (Shackel, 1984). At the true dew point, the diffusive error is minimized.

PELTIER PSYCHROMETERS

Figure 3.6C shows that in the typical configuration for a Peltier psychrometer (Spanner, 1951), Peltier cooling is used to condense water from the vapor atmosphere onto the thermocouple. The cooling is brought about by inserting a battery into the thermocouple circuit and disconnecting the amplifier and readout device. The battery moves an electrical current through the thermocouple in a direction that causes the junction to cool (the Peltier effect). The temperature drops below

---

atmosphere and tissue. No calibration is required. B) Dew point psychrometer using pulsed current from a voltage source to cool the junction to the dew point (Peltier effect). The pulses are varied in size to hold the junction at the dew point temperature. The output of the thermocouple is measured between pulses and compared to outputs previously measured with calibrating solutions. C) Peltier psychrometer using current imposed by a voltage source to cool the junction below the dew point (Peltier effect). Turning off the cooling current allows condensate to evaporate from the junction. The output decreases to a semi-stable value that is compared to outputs previously measured with calibrating solutions.

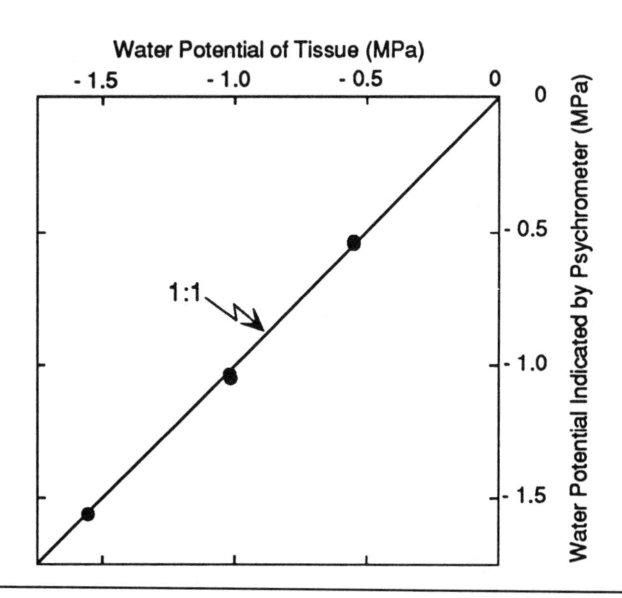

Figure 3.7. Test of the absolute accuracy of an isopiestic psychrometer. Water potential was measured with the psychrometer and compared with the known water potential for the same tissue. The known water potential was achieved by equilibrating the sunflower leaf tissue with an atmosphere of known water potential. Some replicate data points lie too close to one another to be resolved in this plot. From Boyer (1966).

the dew point and water condenses on the thermocouple junction. When the current is switched off, the condensate evaporates to the atmosphere (Fig. 3.6C). The rate of evaporation is greater when the atmosphere contains less vapor. By quickly switching the amplifier and readout into the circuit, the rate of evaporation can be determined and calibrated with solutions of known vapor pressure (water potential, see Appendix 3.1). Comparing the rate on the calibration curve with that of an unknown allows the water potential of the unknown to be determined (Fig. 3.6C).

A variant of this method places a droplet of water on the thermocouple (Richards and Ogata, 1958). The water is held in place around the junction by a porous ceramic bead or a ring. The output of the thermocouple becomes steady as water evaporates. The output is calibrated in the same way as for a Peltier psychrometer.

Neither of these methods operates at thermodynamic equilibrium because they measure the rate of evaporation rather than the vapor pressure that prevents evaporation. As a consequence, calibration is required and is highly sensitive to a number of factors including temperature. Cool temperatures give less thermocouple output than warm temperatures. For best results, the calibration should be repeated immediately before and after the measurement of a sample. Peltier and Richards/Ogata psychrometers have the advantage of simplicity, but the transient nature of the Peltier measurement causes a significant amount of variation in repeated measurements, and Richards/Ogata psychrometers also are variable. With plant tissue and soils, one can expect these psychrometers to have a reproducibility of about ±0.15 MPa (Savage and Cass, 1984; Savage *et al.*, 1983). In both psychrometers, the diffusion characteristics of the vapor path between the thermocouple and the sample affect the measurements (see Diffusion Error), causing systematic error that is usually 5 to 10% of the reading (Boyer and Knipling, 1965) but can be as high as 30 to 50%.

How to Make Measurements

In the following section we briefly discuss the principles for making psychrometer measurements but distinguish between calibrated and noncalibrated instruments. You will need a set of solutions having various water potentials (osmotic potentials), and Appendix 3.2 gives the water potential (osmotic potential) for sucrose solutions. KCl or NaCl solutions also can be used but should be left in contact with the instrument only briefly to prevent corrosion.

PRELIMINARY CHECKS

Electrical Performance. Before using the instrument for the first time, check its electrical performance. Make sure that each thermocouple communicates with the readout (temperature change at thermocouple causes change at readout). Then check that each thermocouple gives a stable reading over long periods of time. It is best to use a recorder to conduct these tests. If there is instability, determine whether it is caused by a loose connection (usually rapid noise) or thermally active junctions (slow drifts in output). Wiggle suspected junctions to check for good contact and warm suspicious parts of the circuit to test for thermal activity. The output of the thermocouples is about 5.0 μV per MPa. Judge the stability in light of the variation you can accept in your final measurements.

For calibrated instruments, check that the thermocouple cooling works properly. Place a solution in the vapor chamber, pass a Peltier current through the thermocouple, and observe thermocouple output immediately after the current is switched off. You should see an output

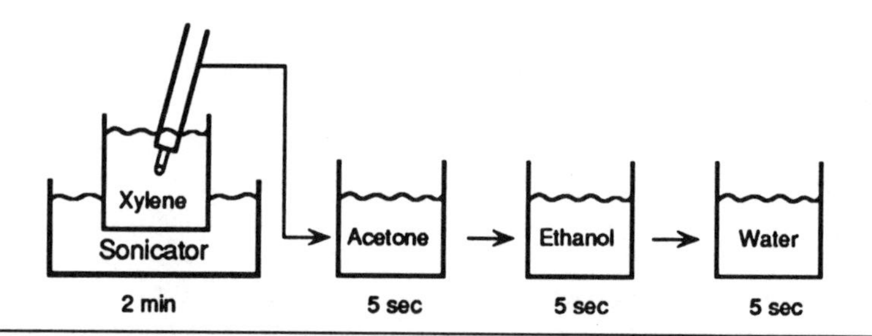

Figure 3.8. Cleaning a thermocouple from an isopiestic psychrometer.

indicating that the thermocouple is cool but the thermocouple should quickly return to zero output.

For isopiestic instruments, check that the thermocouple output is zero when water is on the thermocouple and on the bottom of the psychrometer chamber (Ψ_w = 0). This test is important not only to check whether the thermocouple works isopiestically, but also whether the thermocouple is clean, the electrical circuit is noise-free, and the chamber has been properly loaded with sample. If your thermocouple passes this test, it should be ready for use. Careful handling of the thermocouple usually will preserve this condition for many months, with only occasional testing necessary.

Cleaning the Thermocouples. If the thermocouple is contaminated with foreign material (e.g., Vaseline), the output will drift. Cleaning usually can be done by rinsing in water and blowing dry, but if this does not cure the problem, remove excess foreign material using a small spatula. Submerge the entire thermocouple, including the copper wires at the base, into xylene and sonicate for 1 to 2 min (Fig. 3.8). Follow this with a 5-sec rinse in acetone, then ethanol and finally water. Be careful not to bend the thermocouples during this operation. Reapply fresh Vaseline to the surfaces around the thermocouple as shown in Fig. 3.5, taking care not to cover the thermocouple itself.

ROUTINE PROCEDURES

The vapor pressure of water in samples is so high that surfaces can interact with the vapor and affect the measurements. To avoid this effect, keep chamber surfaces clean and make them as inactive as possible by coating with petrolatum (Vaseline). The following procedures work well for making measurements:

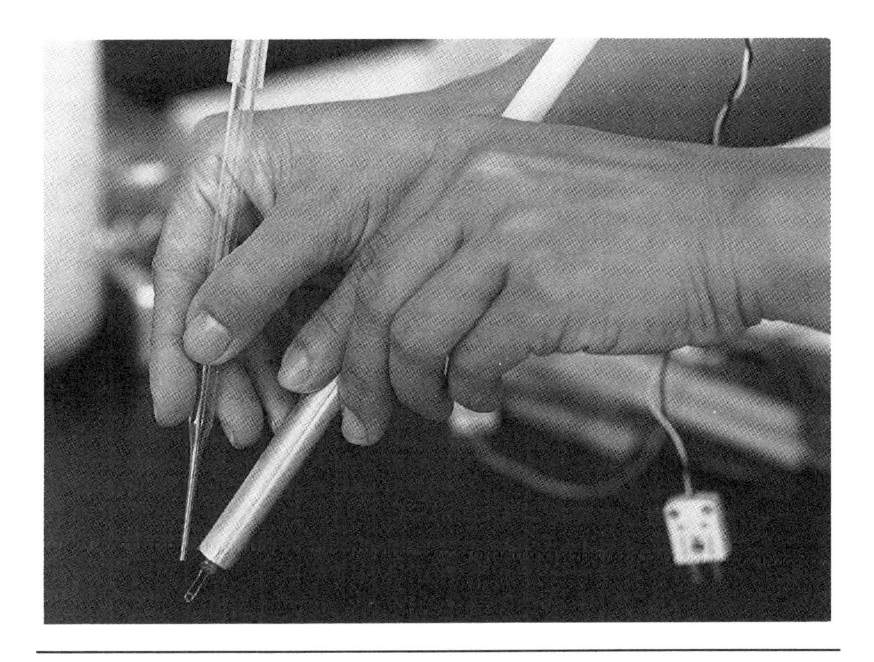

Figure 3.9. Drying the thermocouple from an isopiestic psychrometer in an air stream after rinsing the junction with water. Thermocouple and air stream should be directed downward to move water droplets away from the thermocouple.

1) Clean the thermocouple with a water rinse. Dry in an air stream (Fig. 3.9). Check that all seals are in good condition so that the vapor chamber can be made airtight.
2) Coat the vapor chamber with Vaseline by warming the cup to melt the Vaseline, then covering the bottom and walls and inverting to cool (Fig. 3.10). Use a spatula to spread a thin coating of Vaseline on as much of the chamber top as possible without coating the thermocouple (Fig. 3.5). For isopiestic psychrometers, coat the hole through which the thermocouple enters the chamber by wiping Vaseline into the hole, then removing the excess with a spatula that leaves only a thin coating.
3) For calibrated instruments, calibrate by placing thin tissue paper, i.e., Kimwipe or Kleenex, on the bottom of the cup, and cover the paper with calibrating solution. Assemble the apparatus and note the output of the thermocouple when vapor conditions become stable. Stability usually is achieved within 10 to 20 min. A calibration curve should be

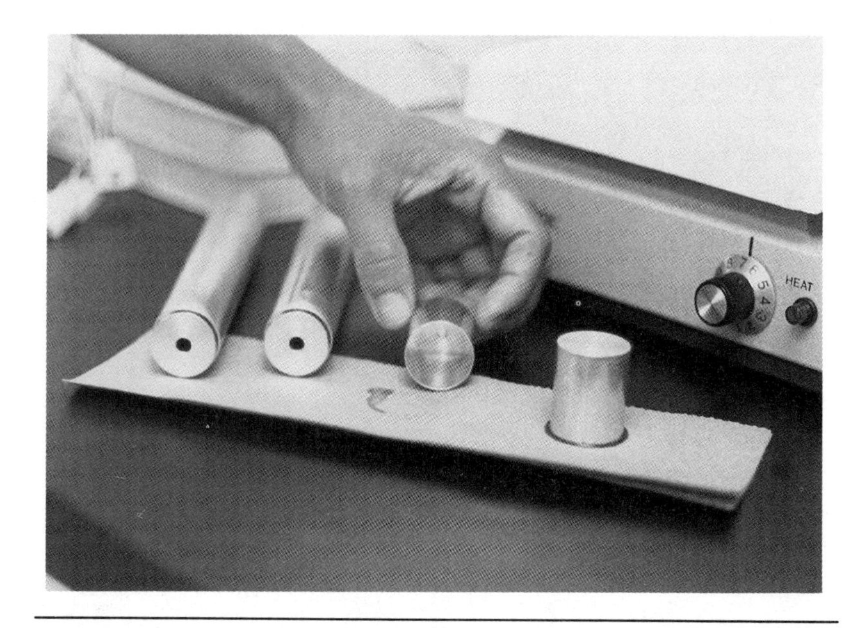

Figure 3.10. Coating the walls of a psychrometer cup with melted and resolidified Vaseline. The cup is cooled completely before reassembling the psychrometer. The chamber top, the walls of the entrance hole for the thermocouple, and the reference junctions of the thermocouple also are coated using unmelted Vaseline as in Fig. 3.5.

constructed to give the output of the system as a function of the water potential (osmotic potential) of the solution as in Fig. 3.11. In general, calibrate before and after measuring each unknown at the temperature being used for the unknown. For isopiestic instruments, this step is omitted.

4) Prepare the tissue or soil to be sampled. The aboveground parts of plants usually have foreign material on the surface. Depending on the growth environment, this may range from dust and salt to spray residues and deposits from watering. Rinse the tissue with water (Fig. 3.12) by wiping with saturated tissue paper, e.g., Kimwipe. Be sure that all surfaces have been thoroughly wetted. Dry by blotting the tissue to remove adhering water. Restore water potential gradients to normal by allowing the tissue to remain in the growth environment for several hours after washing and blotting dry.

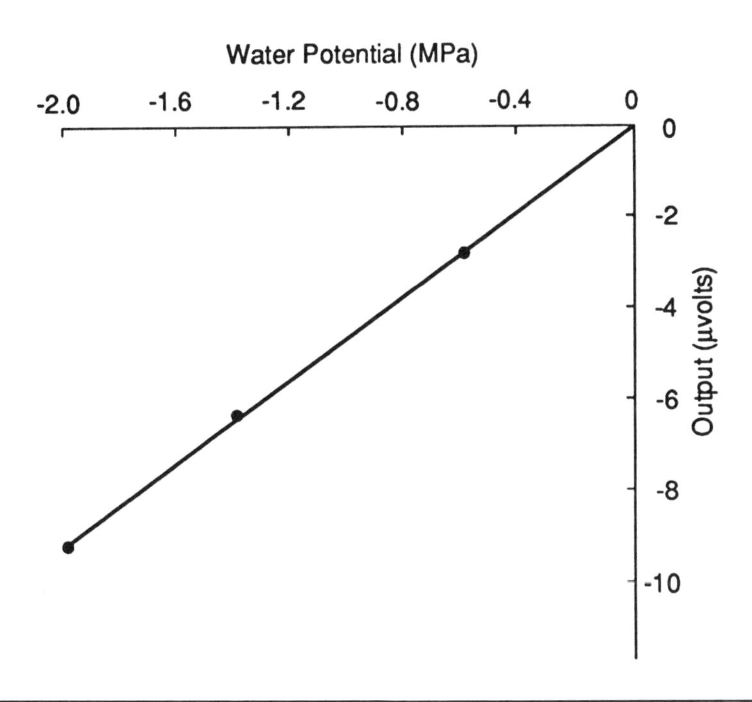

Figure 3.11. Typical calibration curve for Peltier or dew point psychrometer.

Some investigators abrade the surface of plant tissue with carborundum or remove the cuticle with xylene (e.g., Savage *et al.*, 1984). These practices should be avoided because they wound the tissue and disrupt the metabolic and growth activity. Cells lose turgor when they are wounded and they stop growing. Large changes in metabolism can be observed with these treatments. The best measurements are obtainable with clean but otherwise unaltered tissue. Isopiestic psychrometers overcome most of the problems of the cuticle without abrading (see Diffusion Error).

5) Rapidly load the sample into the cleaned and Vaseline-coated cup (see Working with Plant Tissue or Working with Soils for sample placement and handling techniques). Press the tissue into the Vaseline layer at two or three places to ensure good thermal drainage of metabolic heat to the chamber walls. Loading and sealing the cup should take no longer than 10 sec to avoid dehydrating the tissue. For

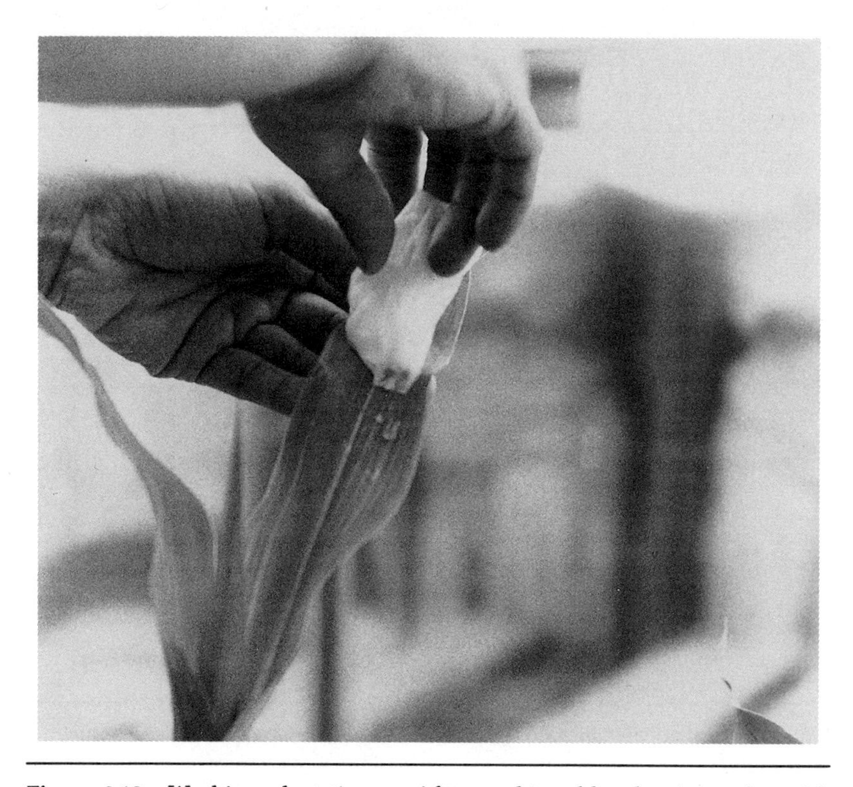

Figure 3.12. Washing plant tissue. After washing, blot the tissue dry with tissue paper. Allow the tissue to reestablish water potential gradients for a few hours before sampling.

leaves, a punch can help because it loads the cup when the disk is cut (Fig. 3.13). If longer sampling times are required, work in a saturated atmosphere. High humidity can be achieved by constructing a simple glove box (Fig. 3.14). Load the cup with one layer of tissue only. Additional layers cause problems because they cannot lose metabolic heat rapidly enough, and liquid condenses between the warm inner layers and cool outer layers causing erroneous readings.

6) Assemble the apparatus. If storage is necessary during transportation from the field, place the apparatus in a Styrofoam box away from direct sunlight, generally for no longer than 15 to 30 min.

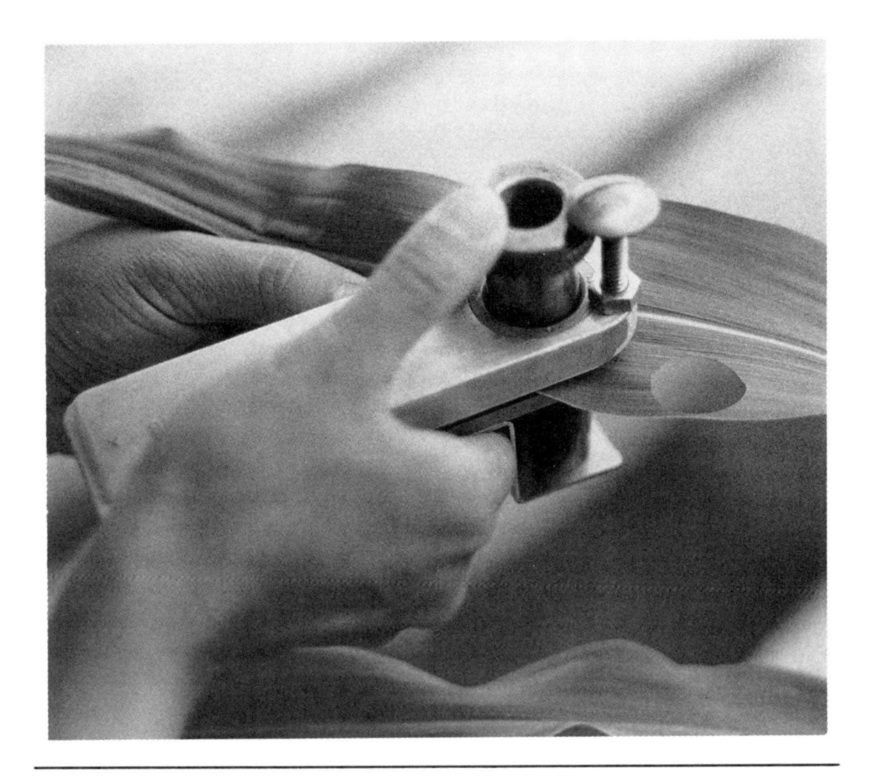

Figure 3.13. Cutting a tissue sample with a leaf punch and loading the psychrometer cup. Work swiftly to avoid sample dehydration.

7) For isopiestic instruments, load water or a solution on the thermocouple. Allow the vapor atmosphere to become stable (0.02 MPa/hr or less). This usually takes 1 to 3 hr, but always test stability by measuring thermocouple output. If times longer than 6 to 8 hr are required, there is usually a problem with the sample. The tissue surface may be dirty (sometimes the tissue may need to be soaked in water for several hours in order to clean the surface, see Step 4 above) or the tissue may degrade during the measurement (rapidly metabolizing tissue in small vapor chambers may run out of oxygen and you will need to use a larger chamber, e.g., 5-10 cm^3, see section on Volume of Vapor Chamber). Some tissues may exude cell liquids onto the surface at high humidities (in this

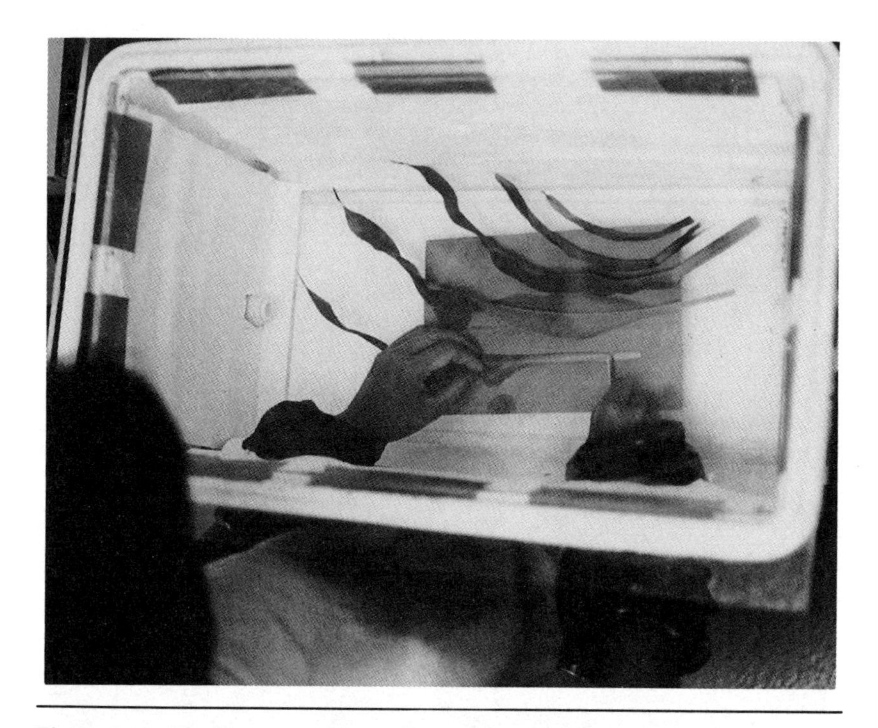

Figure 3.14. Working in an atmosphere saturated with water vapor in a glove box.

case, measurements may not be possible). Occasionally, too little tissue is loaded to adequately communicate with the thermocouple (use more tissue, see section on Working with Plant Tissue).

8) For calibrated instruments, measure the difference between the dry reading before cooling and the stable reading after cooling. Compare these with those during calibration (done with a series of solutions of differing potential prior to the measurements with the tissue) and obtain the water potential of the sample. For the best measurements, it is often necessary to repeat the calibration after the sample measurements have been completed. Calibrations must be at the same temperature as the sample measurements or must be corrected for temperature because these are not equilibrium measurements.

For isopiestic instruments, the calibrations are omitted and the measurements are continued as follows. Remove the thermocouple and replace it with a plunger to seal the psychrometer chamber. Wash the

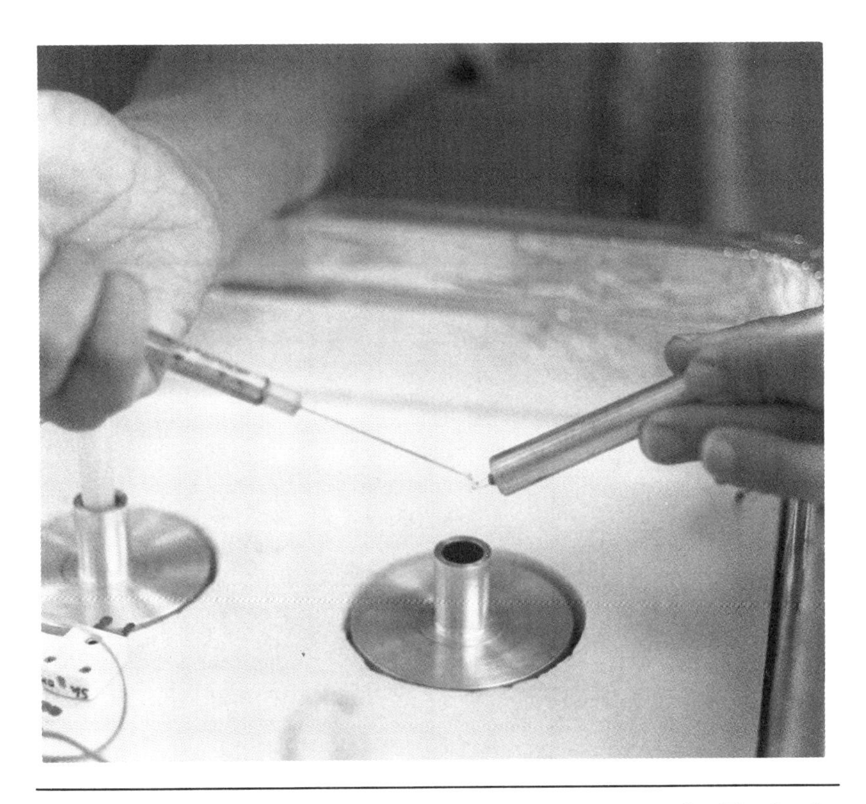

Figure 3.15. Replacing the solution on an isopiestic thermocouple. The droplet size has little effect on the thermocouple output (see Fig. 3.29).

thermocouple in distilled water and dry it with an air jet. Remove the plunger seal, load the spiral with a new solution having a different water potential (Fig. 3.15), and rapidly insert the thermocouple into the psychrometer chamber (Fig. 3.16). Note the new steady output (usually after 0.5 hr, Fig. 3.17), remove the thermocouple, wash and dry it, and insert the dry thermocouple into the chamber. Note the steady output of the dry thermocouple. Because the dry thermocouple neither evaporates nor condenses water from the atmosphere, it identifies the output for the isopiestic solution and corrects for any small thermal activity in the tissue (see section on Isothermal Conditions). Calculate the water potential as shown below. Temperature corrections are unnecessary as long as the potentials chosen from Appendix 3.2 are at the measurement temperature.

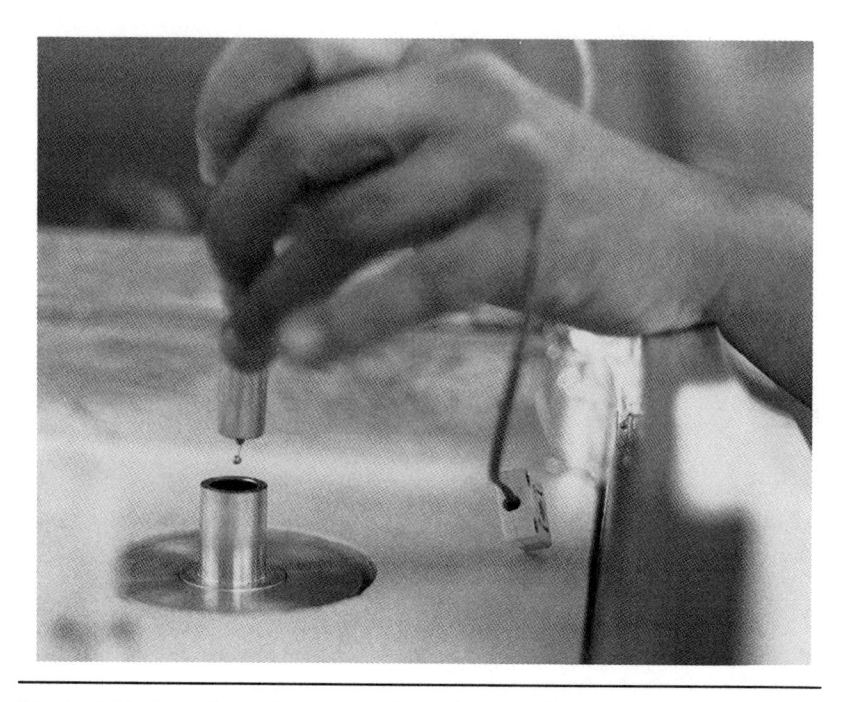

Figure 3.16. Inserting a thermocouple with new solution into a psychrometer system. Work swiftly to avoid dehydration of the droplet.

9) After these steps are completed, remove the sample, clean the vapor chamber and thermocouple junction with distilled water, and dry them with an air jet. Reassemble the unit to protect from dust and damage during storage. This completes the measurements.

CALCULATING WATER POTENTIALS

With Peltier and dew point psychrometers, the water potential is read from the previously prepared calibration curve (Fig. 3.11) by comparing the voltage produced by the unknown with that of the calibrating solutions. For isopiestic psychrometers, the water potential must be calculated by extrapolating to the solution that causes an output that is the same as for the dry thermocouple. The extrapolation is linear because the steady output of the thermocouple is proportional to the potential difference between the thermocouple solution and the tissue (see Fig. 3.18 and also Eq. 3.10). Thus, if a potential difference of 0.5 MPa gives an output of 2 μV, a 1.0 MPa difference will give an output of 4

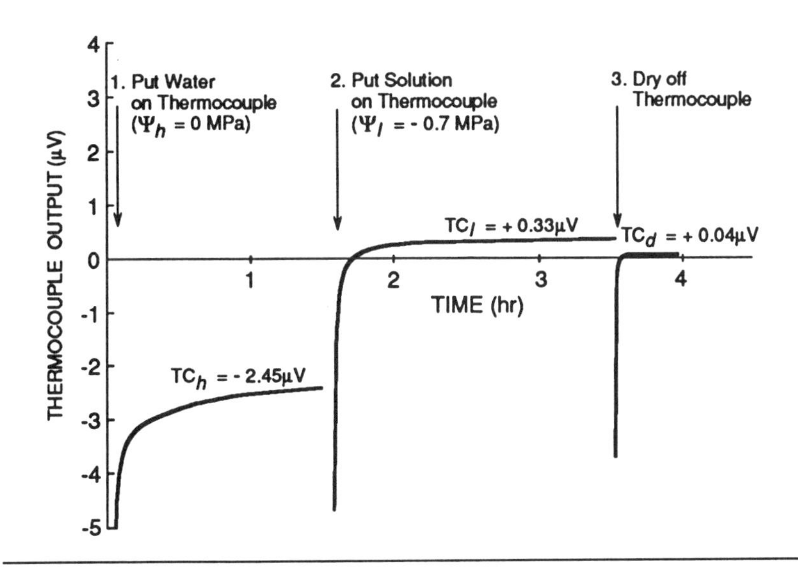

Figure 3.17. Recorder tracing of an isopiestic determination for a sunflower leaf. Measurement 1 gives a rough estimate of the water potential. Measurement 2 gives a more accurate estimate. The isopiestic value is the most accurate estimate and was -0.63 MPa, calculated to give the output of TC_d (Fig. 3.18 and Eq. 3.2).

μV. The isopiestic point is obtained by extrapolating to any fraction of this difference.

The calculation can be formalized with the following equation. Let TC_h be the steady voltage displayed by the thermocouple with the known solution at the higher potential (Ψ_h, closest to 0), TC_l be the steady voltage displayed by the thermocouple with the solution at the lower potential (Ψ_l), and TC_d be the steady voltage of the dry thermocouple (Fig. 3.17). The solution that gives the output of the dry thermocouple is the isopiestic value (Fig. 3.18) and is calculated from:

$$\frac{TC_h - TC_d}{TC_h - TC_l} \cdot (\Psi_l - \Psi_h) + \Psi_h = \text{isopiestic value.} \qquad (3.2)$$

Care should be taken to follow algebraic sign conventions when making this calculation. Thus, a TC_h of -2.45 units (i.e., measurement junction cool) and TC_l of +0.33 units (i.e., measurement junction warm) for solutions

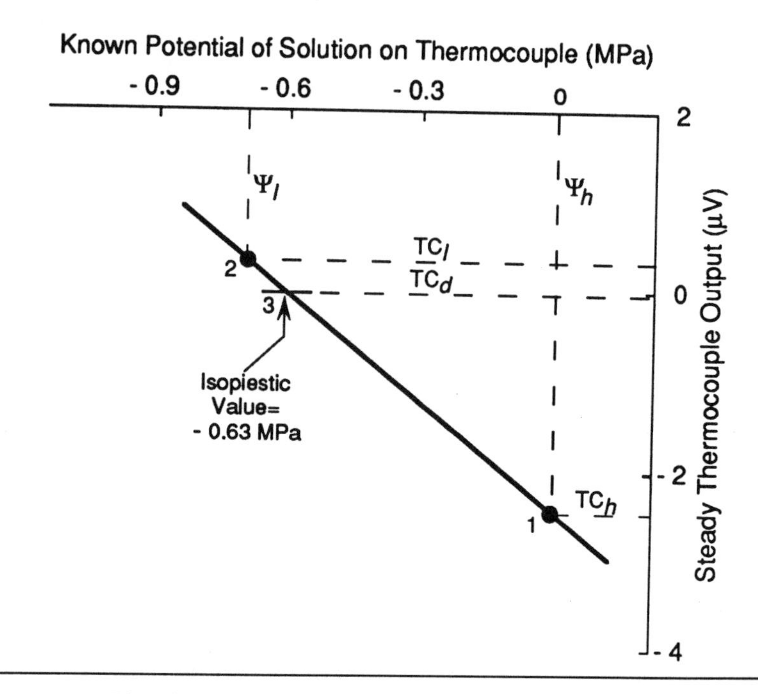

Figure 3.18. Plot of the steady readings in Fig. 3.17. In the first one, water was used (point #1). In the second, a solution was used (point #2). In the third, the thermocouple was dry (point #3). The steady output of the thermocouple was linearly proportional to the difference in potential between the thermocouple and the sample. Therefore, the output could be extrapolated short distances to the isopiestic value as shown (from point #2 to point #3). The extrapolation identifies the solution that would not exchange vapor with the surroundings (that is, the solution causing the same output as the dry thermocouple).

having Ψ of 0 and -0.7 MPa, respectively, with a TC_d of +0.04 units will give an isopiestic value of:

$$\frac{-2.45 - [+0.04]}{-2.45 - [+0.33]} \cdot (-0.7 - 0.0) + 0.0 = -0.63 \text{ MPa.} \qquad (3.3)$$

For bars, multiply this result by 10.

The extrapolation can be checked by placing a solution having the calculated potential on the thermocouple. The reading should be identical to that of the same thermocouple when dry.

Working with Plant Tissue

Because thermocouple psychrometers detect the vapor pressure of the surface water, the vapor must proceed from the outer surface of the cells through the intercellular spaces to the atmosphere to be detected by the thermocouple. However, at equilibrium in excised tissue, the water potential of the cell interior is in equilibrium with the water potential at the cell surface that in turn is in equilibrium with the atmosphere in the psychrometer chamber. There is no net water or vapor movement and the entire system is at the same Ψ_w.

In intact plants, *in situ* measurements allow gradients in water potential to exist over distances of several cells because the tissue remains attached to its water supply and water movement can occur. However, water vapor diffuses rapidly and the vapor usually represents an average for the gradient. Boyer *et al.* (1985) showed that psychrometers give average values for the sample whether water is entering or not and thus whether gradients are present or not. This simplifies the interpretation of psychrometer measurements.

It is worth considering what kind of average the psychrometer provides. As discussed in Chap. 2, excising plant tissue removes its water supply, and preventing evaporation allows internal gradients in water potential to equilibrate, giving an average for the tissue at the time of excision. Tyree and Hammel (1972) and Tyree and Jarvis (1982) point out that at equilibrium the potential for a tissue should be a volume average that accounts for the fact that the volume of water having a particular potential determines the amount of work that can be done by that water, and larger volumes will do more work and thus contribute more to the average than small volumes will. Accordingly, the volume-averaged Ψ_w is

$$\text{Average } \Psi_w = \frac{\Sigma\, V^i \cdot \Psi_w^i}{V}, \tag{3.4}$$

where V^i is the water volume in the protoplasm of cell i, Ψ_w^i is the water potential of cell i, and V is the total water volume in the protoplasm of all the cells (the symplasm). The symbol Σ sums the effects of each cell V^i and Ψ_w^i in the tissue, and the $V^i \cdot \Psi_w^i$ weights the potential of each cell by the volume of water in that cell. This volume-averaging concept applies to the components of Ψ as well as to any other cell parameters that depend on tissue measurements of Ψ.

Figure 3.19 shows the effect of volume averaging on the Ψ_w measured in a leaf. Before excision, there is a gradient in Ψ_w extending

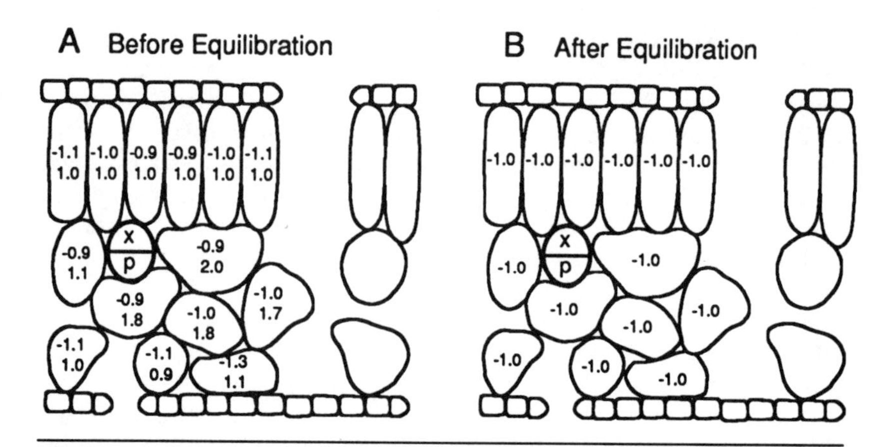

Figure 3.19. Water potential gradients in a hypothetical leaf before and after equilibration in a vapor chamber. A) Before equilibration, a gradient exists in the intact, freely transpiring leaf. The Ψ_w is highest (-0.9 MPa) next to the vein (x/p) and lowest (-1.3 MPa) far from the vein (see negative values in each cell). The volume of water in each cell is also shown (positive values). B) After the sample is excised and equilibrated internally, the Ψ_w represents a volume average for the gradient. The volumes change slightly in each cell as water is exchanged between the cells during the approach to uniform potentials. Regardless of whether a gradient exists as in A or equilibration has occurred as in B, vapor pressure measurements with psychrometers are volume averages because the vapor mixes rapidly in the intercellular spaces,

from the veins to the leaf surface. The Ψ_w are higher close to the veins (-0.9 MPa) than far from the veins (-1.3 MPa), and the volume of water varies from cell to cell (e.g., see Nonami *et al.*, 1991). After excision, the Ψ_w equilibrate according to Eq. 3.4, and the Ψ_w is -1.0 MPa in all the cells.

A similar principle holds for the vapor above the gradient. The vapor pressure above cell i reflects Ψ_w^i (Eq. 3.4) and, because the vapor in the intercellular spaces is very mobile compared to the liquid, the vapor mixes in a volume-averaged fashion. Thus, the psychrometer gives a volume-averaged Ψ_w whether a gradient is present or not (Boyer *et al.*, 1985).

From the foregoing, it is clear that two approaches may be taken to ensure that the sample represents the potential in the intact plant. In the first, one may remove the sample from the plant and rapidly seal it in the vapor chamber so that internal equilibrium occurs and the measurement represents a volume average of the potentials present in the intact plant before excision. Alternately, one may enclose part of the intact plant in a vapor chamber and measure the volume average of any gradients that are present. Although gradients may persist, the volume average clearly applies to the intact plant. However, by enclosing the tissue, transpiration is prevented and the potentials in the intact plant may not be the same as before enclosure. The excised tissue has the advantage that the volume average applies to the Ψ_w before excision but the intact plant has the advantage that no excision artifacts are possible. Depending on the experiment, one approach may be more desirable than the other.

EXCISED TISSUE

Because the major value of excised tissue is that it indicates the average water status at the instant of excision in the freely transpiring plant, it is particularly useful for routine measurements with a wide range of tissues. As long as the ratio of cut surface to intact surface is small (generally less than 10%), the cut surface does not affect the psychrometer significantly and can be neglected. If the cut surface is more extensive, it should be hidden by coating with Vaseline (see section on Sorption Effects).

For relatively dry plant tissue or soil where the water content is small, it is important to avoid dehydrating or rehydrating the tissue with the water on the thermocouple. Isopiestic psychrometers use potentials close to those of the tissue (thermocouple output is kept small), and hydration changes can be kept minimal. Peltier and dew point systems use water from the tissue or soil to humidify the air and coat the thermocouple, so a certain amount of dehydration is inevitable.

Leaves. After cleaning the leaf surfaces with water (Fig. 3.12), blotting dry, and allowing the leaves to reequilibrate with their surroundings for several hours, excise a sample using a leaf punch (Fig. 3.13). Place the disk on the bottom (Fig. 3.2) of the Vaseline-coated psychrometer cup and seal it within 10 sec of removal from the plant. If no leaf punch is available, place an oversized leaf fragment in a glove box having a saturated atmosphere (Fig. 3.14) and trim a sample to size. For needles or small fragments of leaves, work in the glove box and place the tissue fragments on the walls and the bottom of the cup until an area equivalent to the bottom is covered. Press the tissue into the Vaseline on the cup surfaces to hold it in position and to drain away metabolic heat. Use only one layer of tissue and do not allow any overlap between the fragments.

Leaf disks with diameters of 2.0 to 2.5 cm usually are large enough to have negligible cut surfaces. Because it is desirable to avoid the effects

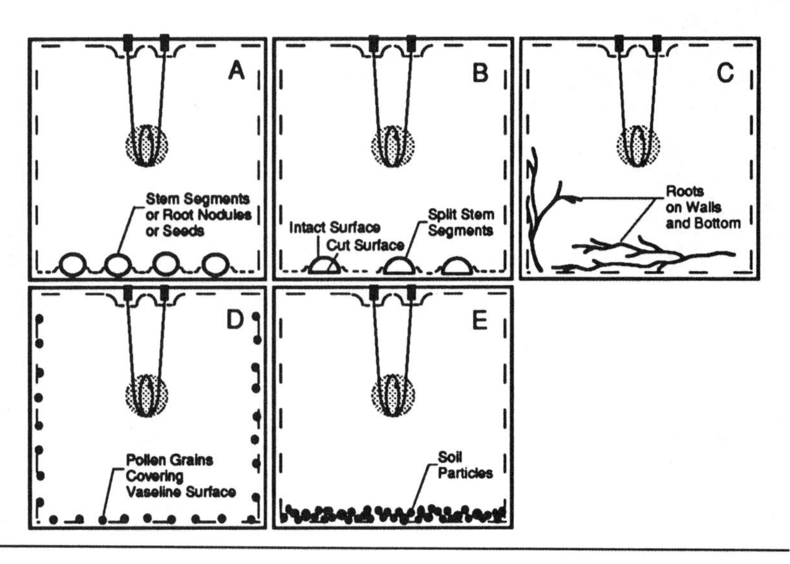

Figure 3.20. Placement of various plant parts or soil in the vapor chamber. A) Whole stem segments, root nodules, or seeds. B) Split segments of large stems. C) Roots. D) Pollen grains. E) Soil. Leaf placement is shown in Fig. 3.2.

of cut surfaces as much as possible (Nelsen *et al.*, 1978), leaf samples should be this size or larger. If thick leaves are being sampled, large amounts of cut surface may be inevitable. In this case, coat the edge with Vaseline before loading.

Stems. Place the stems into a saturated glove box (Fig. 3.14) for sampling. Small diameter stems (to 3 mm) can be cut into lengths of 2.0 to 2.5 cm without exposing too much cut surface. Place four to six segments in the cup so they do not touch each other (Fig. 3.20A). For larger stems, cut short lengths and split the stems. With the split surface down, submerge the cut surfaces in the Vaseline on the bottom (Fig. 3.20B). The Vaseline layer should be thick enough to completely bury the cut surface, hiding it from the vapor atmosphere. Of the surface exposed to the thermocouple, at least 90% should consist of intact cells.

Roots. Place the soil/root complex into a saturated glove box (Fig. 3.14) and dissect the root segment to be measured. Flick away adhering soil particles. Use caution because even light brushing can cause the cells to collapse. For short segments of small diameter (to 3 mm), place four to six segments on the cup bottom and the walls (Fig. 3.20C). The cut surface generally can be neglected. Avoid root-root contact between the samples. For long roots, wrap the root around your index finger and

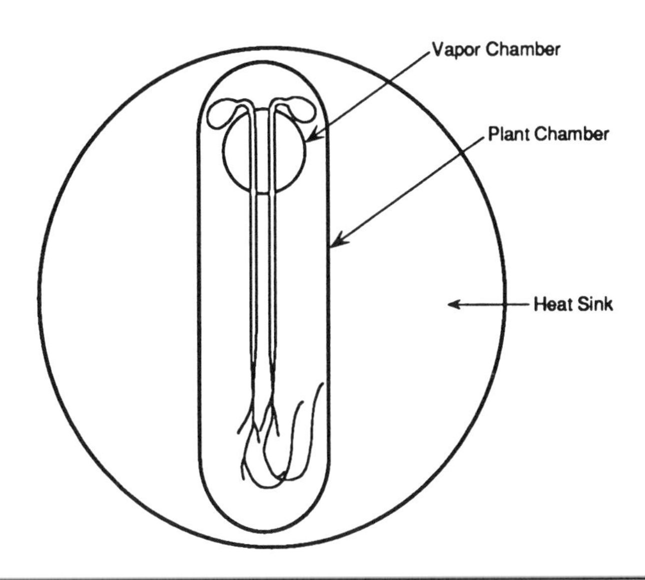

Figure 3.21. Example of a thermocouple psychrometer for measuring the water potential in intact plants. Etiolated seedlings are held in a plant chamber in a large aluminum heat sink. The vapor chamber for the psychrometer encloses the part of the plant to be measured. The vapor chamber is sealed around the stems with Vaseline.

insert it into the cup, releasing it so the root can uncoil against the wall. Check to be certain that no roots extend into the region to be occupied by the thermocouple. For fleshy roots, cut a section but coat the cut surfaces with Vaseline and expose the remaining uncoated surface to the thermocouple. At least 90% of the exposed surface should be intact.

Root Nodules. Place six to eight small nodules or three to four large nodules into the cup and press into the Vaseline on the bottom (Fig. 3.20A). The Vaseline layer should be thick enough to cover the lower half of the nodules and to drain away metabolic heat. Cut surface is usually negligible.

Seeds. Same as with nodules (Fig. 3.20A).

Pollen. Collect pollen in a glassine bag. Rapidly insert the bag into a saturated glove box (Fig. 3.14) and cut away the lower corner of the bag. Pour the pollen from the open corner into the cup. Tilt the cup to coat the Vaseline-covered surface with pollen (Fig. 3.20D). Place the cup on the thermocouple unit.

INTACT TISSUE

The psychrometer is the only method of measuring plant water status in completely intact plants. One advantage of using intact tissue is that it can be observed for long times while growth occurs or the tissue hydrates (Boyer *et al.*, 1985). This is because the attachment to the rest of the plant gives a path along which oxygen, water, and nutrients can move and keep the tissue vigorous.

Measurements of intact tissue follow the same principles as for excised tissue and are made by sealing the tissue into an airtight, temperature-controlled vapor chamber. The measurements require specialized instruments, an example of which is shown in Fig. 3.21. As in any other psychrometer measurement, it is essential to use clean tissue, coat chamber surfaces with melted and resolidified Vaseline, and maintain good contact between the tissue and the Vaseline to drain away metabolic heat.

Working with Soils

SOIL SAMPLES

The water potential of soils can be measured with samples placed in the psychrometer cup. Put an oversized soil sample into a saturated glove box (Fig. 3.14). Subsample the soil sufficiently to cover the bottom of the cup with a layer 2 or 3 mm thick (Fig. 3.20E). An approach that has worked well is to obtain a soil core using a tube with a sharpened edge, then push out the core and catch a sample from mid-core in the psychrometer cup. Be careful to avoid regions of soil that have been subject to air drying because they usually do not represent the bulk environment around the roots. When placing the cup on the thermocouple unit, keep it upright to prevent soil from making contact with the thermocouple. For relatively dry soils where water films are small, avoid large potential differences between the soil and the thermocouple. Large potential differences allow too much vapor transfer between the soil and the thermocouple, which can hydrate or dehydrate the soil and prevent a steady output from being achieved. To keep potential differences small with an isopiestic psychrometer, place a solution on the thermocouple that has a potential close to that of the soil. With extremely dehydrated soil, it may be necessary to operate the psychrometer in an atmosphere of high humidity to avoid dehydrating the sample when changing solutions during a determination. For Peltier and dew point instruments, some dehydration of the sample is inevitable because of the condensation of water from the sample onto the thermocouple surface.

In Situ SOIL MEASUREMENTS

Some psychrometers are built to be buried in soil where they can provide frequent measurements at constant positions in the soil profile (Brown and Collins, 1980). The thermocouples should be calibrated before being placed in the soil. The procedure is to expose the thermocouples to the atmosphere above salt solutions of known water potential in a temperature-controlled water bath. After calibration, bury the thermocouples and the surrounding protective housing, such as a porous ceramic cup or screen cage, well before measurements are taken so that the soil has a chance to settle (Brown and van Haveren, 1972). Generally, calibrations should be repeated immediately after the psychrometer is removed from the soil at the end of the experiments. This can detect any shift in calibration that occurred while the thermocouples were in the soil. In theory, isopiestic measurements could be made *in situ* and would avoid the need for calibration, but they have not been attempted.

Measuring the Components of the Water Potential

So far this chapter has dealt with measuring the water potential, but it is often just as important to determine the components of the water potential (Eq. 2.1). Because psychrometry indicates the vapor pressure of the surface solution in the walls of the cells, the potentials in other parts of the cell are measured by appropriate mixing with the surface solution. As pointed out earlier, the surface solution in the apoplast consists of the components

$$\Psi_{w(a)} = \Psi_{s(a)} + \Psi_{m(a)}, \tag{3.5}$$

and in the protoplasts there are different components

$$\Psi_{w(p)} = \Psi_{s(p)} + \Psi_{p(p)}. \tag{3.6}$$

Each equation contains three variables and measuring any two allows the third to be calculated. A simplification is the equilibrium between the wall solution and the protoplast solution that allows $\Psi_{w(p)}$ to be considered identical to $\Psi_{w(a)}$. Measuring the osmotic potential in both compartments then allows the other component ($\Psi_{m(a)}$ or $\Psi_{p(p)}$) to be calculated from Eqs. 3.5 and 3.6.

OSMOTIC POTENTIAL

Solutions sometimes behave nonideally because of dissociation of the solute (NaCl for example) or binding of water to the solute (some sugars and macromolecules) or relatively independent motions of various parts of polymers. These effects alter the Ψ_s of the solution from that expected from the concentration of solute alone. Thermocouple psychrometers measure water activities in the vapor phase and thus in the liquid phase. Any nonideal effects that alter liquid water activity are detected and the true Ψ_s is automatically measured.

Apoplast Osmotic Potential. The $\Psi_{s(a)}$ can be determined as described in Chap. 2 by using a pressure chamber to obtain a sample of apoplast solution (Scholander *et al.*, 1964, 1965; Boyer, 1967a; Nonami and Boyer, 1987; Jachetta *et al.*, 1986).

1) Excise a leaf, branch, or root system and place it in the pressure chamber so that the cut end extends through the top to the outside of the chamber.
2) Place water on the cut surface and wipe it away several times to remove solutes released from the cut.
3) Pressurize the tissue until apoplast solution appears on the cut surface and wipe away the first 2-5 µl to remove any solute that may have been released during the pressurization.
4) Collect the next 5 µl of exudate in a microliter syringe.
5) Place a droplet of exudate on the spiral thermocouple of an isopiestic psychrometer above a known solution on the bottom of the cup. After the output of the thermocouple becomes steady, move the thermocouple to a second vapor chamber containing a different known solution (Fig. 3.22) and obtain a steady reading. The isopiestic point is found by extrapolation. The calculation is similar to that for measuring Ψ_w with living tissue (Eq. 3.2) except that TC_d is zero (no metabolic heat) and the definition of the other TC terms is *reversed* because the unknown is on the thermocouple rather than in the cup. Accordingly, let TC_h be the steady voltage displayed by the thermocouple with the *cup solution* having the higher potential (Ψ_h, which gives the higher thermocouple voltage), and TC_l be the steady voltage displayed by the thermocouple with the *cup solution* having the lower potential (Ψ_l, which gives the lower thermocouple voltage), and TC_d be zero. Thus, a TC_h of +3.0 units above a solution with Ψ_h of -0.4 MPa and a TC_l of -7.2 units above a solution with Ψ_l of -0.7 MPa will give an osmotic potential (Fig. 3.22) of

$$\frac{+3.0 - 0}{+3.0 - [-7.2]} \cdot (-0.7 - [-0.4]) + (-0.4) = -0.49 \text{ MPa.} \qquad (3.7)$$

This procedure works well for determining $\Psi_{s(a)}$ in samples of 3-5 µl and, by bending the thermocouple spiral to a smaller size, isopiestic psychrometers can measure samples as small as 0.1 µl. It is also possible to presample a leaf, branch, or root system to measure the water potential in a psychrometer and then pressurize the remaining plant part in a

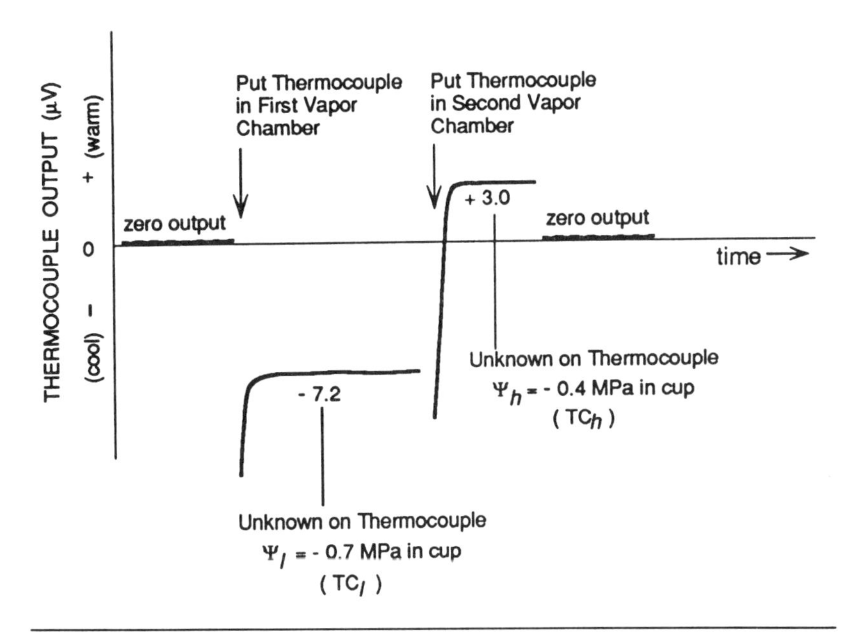

Figure 3.22. Recorder tracing of an isopiestic osmometer measurement. The isopiestic value is -0.49 MPa, which is the osmotic potential needed in the cup to give a thermocouple reading of zero. See Eq. 3.7 for the calculation.

pressure chamber to remove the xylem solution for measuring $\Psi_{s(a)}$. With these presampled tissues, it may be necessary to coat cut surfaces with Vaseline in order to prevent excessive leakage of gas in the pressure chamber.

Occasionally it is possible to obtain apoplast solution from completely uncut tissue by pressurizing the roots of intact plants, collecting exudate from the surfaces of the shoot tissue, and measuring $\Psi_{s(a)}$ of the exudate. This method has proven successful for obtaining apoplast solution from stem elongating tissues (Nonami and Boyer, 1987) and from leaf hydathodes, which exude xylem exudate (Klepper and Kaufmann, 1966). It has been possible to use a microcapillary to sample the region below developing seeds (Maness and McBee, 1986).

Protoplast Osmotic Potential. The $\Psi_{s(p)}$ is obtained by breaking the protoplast membranes, extracting the mixed apoplast/protoplast solution, and measuring the osmotic potential of the solution (Ehlig, 1962).

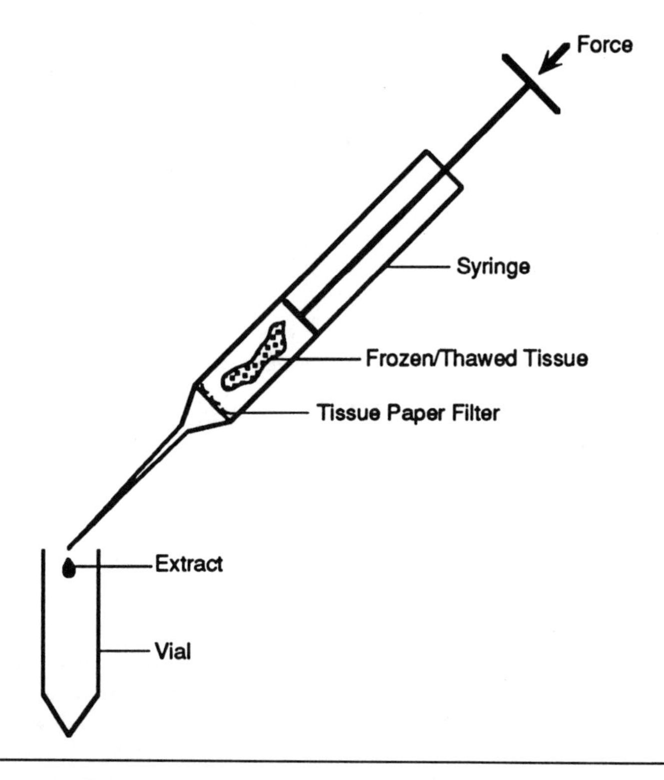

Figure 3.23. Extraction procedure after freezing and thawing plant tissue in the barrel of a syringe. Note tissue paper filter to prevent plant tissue from entering the syringe needle.

1) Place a tissue sample into the barrel of a 1-ml syringe with a half millimeter wad of tissue paper as a filter at the bottom (Fig. 3.23). Insert the plunger above the tissue, insert the needle tip into a rubber stopper to act as a seal, and place the syringe/stopper into a freezer.
2) After at least 20 min in the freezer, remove the syringe and thaw. Immediately after thawing, unseal the needle tip, press firmly on the syringe plunger, and catch the cell extract in a vial as it exudes from the syringe needle. Seal the vial and mix the extract.
3) Place a droplet of the extract on the thermocouple spiral of an isopiestic psychrometer (usually 3-5 µl volume) and measure $\Psi_{s(p)}$ as described earlier for measuring the apoplast osmotic potential. If you have a large volume of extract (usually more than 50 µl), it is possible to use a Peltier

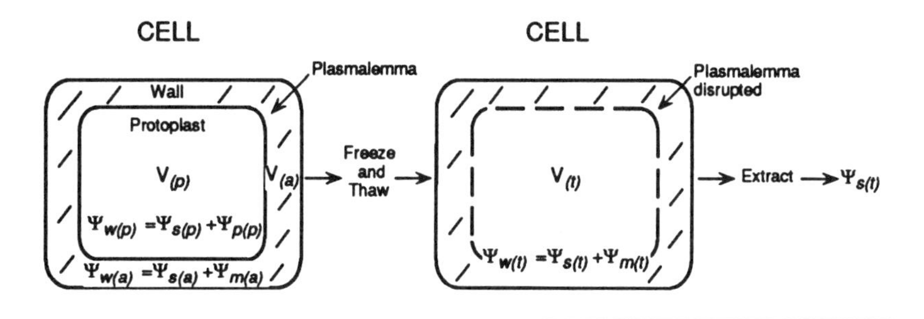

Figure 3.24. Compartmentation of components of the water potential before and after freezing and thawing a tissue sample. Initially, the protoplast solution in the cells is separated from the wall by the plasmalemma. After freezing and thawing, the compartments $V_{(a)}$ and $V_{(p)}$ mix to give a single solution $V_{(t)}$. The $\Psi_{w(t)}$ of the frozen/thawed solution indicates $\Psi_{s(t)}$, which has been diluted by the water in the walls, and $\Psi_{m(t)}$, which has been flooded by solution from the protoplast. The $\Psi_{s(p)}$ can be determined by correcting $\Psi_{s(t)}$ for dilution after extracting the cell solution (see Eq. 3.9). The volume $V_{(t)} = V_{(a)} + V_{(p)}$.

or dew point psychrometer to measure the osmotic potential. Place the extract into the vapor chamber and determine the osmotic potential from a calibration curve made with solutions of known osmotic potential (see Appendix 3.2).

Sometimes it is not possible to make an extract for the measurement of $\Psi_{s(p)}$. An alternate method is to measure the osmotic potential *in situ* in the frozen and thawed tissue. To do this, freeze the tissue in the psychrometer cup after sealing the top with plastic film. Freeze as rapidly as possible, preferably in liquid nitrogen. The faster the freezing rate, the more uniform the ice crystal distribution and the shorter is the time necessary to establish steady state conditions after thawing. Thaw the walls and top of the chamber first to drive any condensed water off these surfaces and onto the tissue. Immediately place the cup onto the thermocouple assembly and measure the water potential of the thawed tissue ($\Psi_{w(t)}$), then measure the matric potential ($\Psi_{m(t)}$) as indicated in the section on Matric Potential (the subscript *t* refers to frozen/thawed tissue). Calculate $\Psi_{s(p)}$ from $\Psi_{w(t)} - \Psi_{m(t)}$.

Compartment Mixing. As long as the wall volume is small, there is only a small error caused by mixing the apoplast and protoplast solutions during thawing. If the wall volume is significant (5-10% of cell volume or greater), mixing causes significant error because the apoplast solution is dilute but the protoplast solution is concentrated (Boyer and Potter, 1973). Consider the volume of water in the protoplasm ($V_{(p)}$) and apoplast ($V_{(a)}$) to be mixed in the total volume after freezing and thawing ($V_{(t)}$; Fig. 3.24). To correct for mixing, assume the apoplast solution is pure water (an oversimplification, but the solution is usually so dilute that the assumption is safe). The protoplast solution mixed with the apoplast solution after freezing and thawing is $\Psi_{s(t)}$. The relationship of $\Psi_{s(p)}$ to $\Psi_{s(t)}$ is then

$$\Psi_{s(p)} \cdot V_{(p)} = \Psi_{s(t)} \cdot V_{(t)} \tag{3.8}$$

and

$$\Psi_{s(p)} = \Psi_{s(t)} \frac{V_{(t)}}{V_{(p)}}. \tag{3.9}$$

For example, $\Psi_{s(t)}$ of -1.0 MPa measured with extract from frozen/thawed tissue having $V_{(t)}/V_{(p)}$ of 1.2 will have a $\Psi_{s(p)}$ of -1.2 MPa. The volume $V_{(t)}$ and $V_{(p)}$ can be measured microscopically in fresh tissue sections (also see Chap. 2 for another method of measuring $V_{(t)}$ and $V_{(p)}$).

TURGOR

The $\Psi_{p(p)}$ is measured by first determining the water potential of the living tissue ($\Psi_{w(a)} = \Psi_{w(p)}$) followed by measuring the $\Psi_{s(p)}$ as above. Calculate the $\Psi_{p(p)}$ by difference according to Eq. 3.6.

MATRIC POTENTIAL

Because matric potentials arise from surface effects of porous solids in contact with water or solutions (Fig. 3.25), the $\Psi_{m(a)}$ of the surface solution can be measured with a psychrometer in living tissue by first determining $\Psi_{w(a)}$ and $\Psi_{s(a)}$ as described earlier, then calculating $\Psi_{m(a)}$ by difference from Eq. 3.5.

As discussed for measuring the osmotic potential of tissues *in situ* after freezing and thawing, it sometimes is desirable to measure the matric potential in frozen/thawed tissue ($\Psi_{m(t)}$). The matric potential is no longer that of the intact tissue (Boyer, 1967b) because the wall has been flooded with solution from the protoplasm, and the matric potential becomes less negative. If you wish to measure $\Psi_{m(t)}$, use an isopiestic psychrometer, and place a solution extracted from frozen/thawed tissue on the thermocouple and a frozen/thawed sample of the same tissue in the cup (Fig. 3.26). The difference in potential between the solution on the thermocouple, which is affected only by the osmotic component, and the frozen/thawed tissue in the cup, which is affected by the osmotic

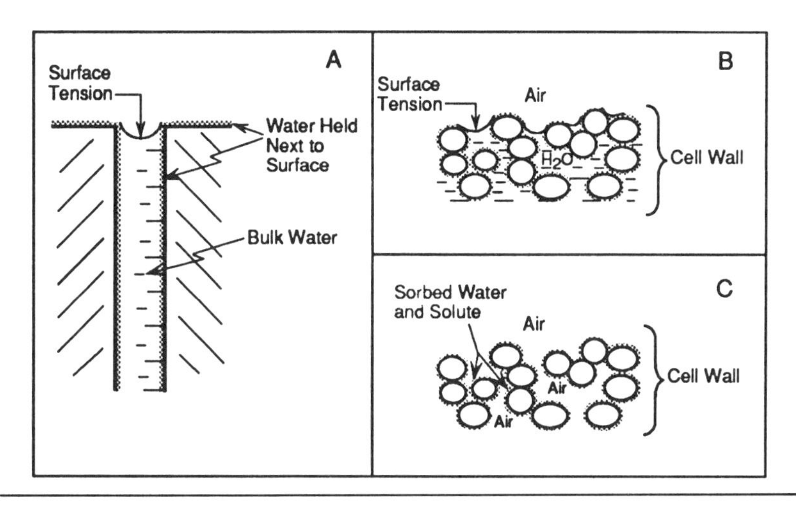

Figure 3.25. Origin of matric potential. A) In a single capillary with completely wettable walls, molecules of water are held next to the wall mostly by hydrogen bonds and electric charges. Electrolytes are attracted by charges on the wall, and the high solute concentrations also attract water. Water tends to fill the pore because hydrogen bonds, electrostatic attractions between the weak water dipoles, and other intermolecular forces hold the liquid water together. There is a meniscus at the air-water interface maintained by the intermolecular forces, which create surface tension. B) In a cell wall, there are many small pores (about 5 to 8 nm diameter) that remain water-filled against large tensions. The round structures in this view are cross sections of the microfibrils and matrix polymers forming the solid structure of the wall. Water fills the spaces and is continuous with the protoplasts and the vascular system, forming a hydraulically connected system. C) In a cell wall that has been desiccated sufficiently to drain the pores, hydraulic contact is lost between the pores and the protoplasts and vascular system. Air fills most of the pores and pressures cannot be measured easily. However, water remains adsorbed to microfibrils and its vapor pressure can be measured with a thermocouple psychrometer.

and matric components, gives the matric component of the frozen/thawed tissue.

GRAVITATIONAL POTENTIAL

Although we often ignore gravitational potentials, they can be substantial. For every 10 m of depth, gravity causes an increased potential of 0.1 MPa. The effects are observed mostly in tall trees (Scholander *et* al., 1965), marine environments, and deep soils. In a column of free water, the gravitational potential is expressed as pressure that increases at

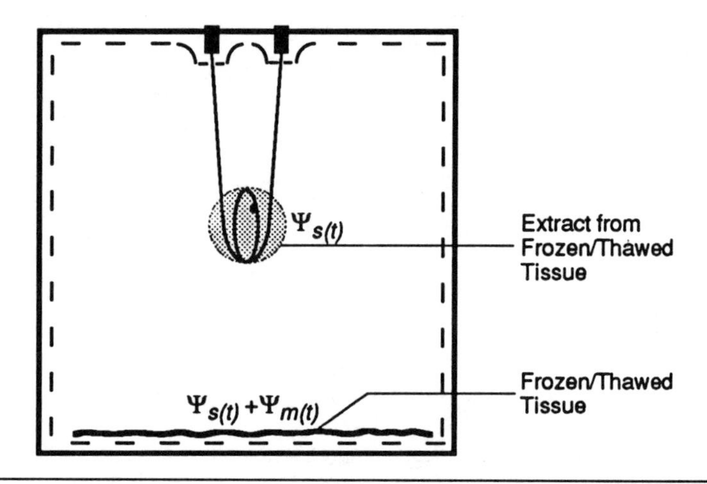

Figure 3.26. Measurement of matric potential in frozen/thawed tissue *in situ* ($\Psi_{m(t)}$). The difference in potential between the frozen/thawed tissue ($\Psi_{s(t)}$ + $\Psi_{m(t)}$) and the osmotic potential ($\Psi_{s(t)}$) extracted from a parallel sample is the $\Psi_{m(t)}$ in the frozen/thawed tissue.

increasing depths. If the column is in a capillary and the weight of water is allowed to put a tension on the meniscus in the capillary, as in a tree, the pressure will be negative immediately under the meniscus and will increase (become less negative) at increasing depths.

The difference in potential at two heights indicates the effect of gravity only if the water column is stationary and the composition of the solution is uniform. In tall trees and in soils, the water often is moving and the solution may not be uniform in composition, so care must be taken to measure all other potentials. Psychrometers can be mounted at various positions and gravitational effects can be monitored *in vivo*. For excised tissue, excision and psychrometer loading should take place at the sampling height in the tree to allow rapid sampling (within 10 sec).

Precautions

DIFFUSION ERROR

Diffusion determines the interaction between the thermocouple and the sample and involves vapor and heat transfer (Boyer, 1969b; Boyer and Knipling, 1965; Rawlins, 1964). The physical arrangement can be idealized to two exchange surfaces, one at the thermocouple and one inside the tissue. For calibrated psychrometers, measurements are not made

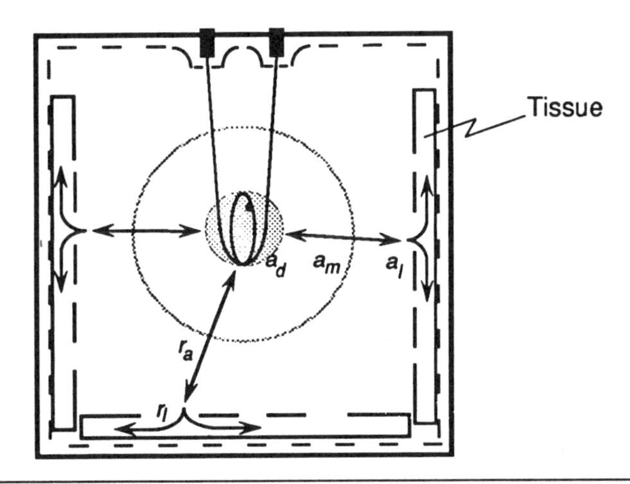

Figure 3.27. Diffusion properties of a thermocouple psychrometer. In order to simplify the mathematics, the tissue lines the sides and bottom of the vapor chamber and is shown with a porous surface that restricts diffusion due to stomata. The diffusive resistance of the air between the thermocouple droplet and tissue (r_a) is in series with the diffusive resistance of the porous tissue (r_l). The area governing diffusion is the geometric mean area (a_m, shown as a shaded sphere around the thermocouple) that accounts for the differing areas of the droplet (a_d) and tissue (a_l).

at equilibrium and water vapor diffuses between the two surfaces governed by the geometry, surface area, and diffusive resistance of each part of the diffusion path. There is always a possibility of diffusion error in these instruments (Boyer and Knipling, 1965; Shackel, 1984). On the other hand, there is no net vapor movement at the isopiestic point in isopiestic psychrometers and thus there is no diffusion error. This difference can have an impact on measurements with psychrometers.

Because of the opportunity for diffusion errors in calibrated instruments, it is useful to analyze vapor transfer and how the errors originate. Figure 3.27 shows the diffusion properties of a psychrometer with a water droplet on the thermocouple and plant tissue lining the walls. There are two paths in series for vapor transfer when tissue is present and each has its own diffusive resistance (Fig. 3.27). One is between the droplet and the sample surface (diffusive resistance r_a, sec·m^{-1}) and the other is between the sample surface and the interior water surface (diffusive resistance r_l). The diffusion surfaces are of area a_d (m^2) for the water drop and a_l for the sample that can be simplified to a single area a_m which is the geometric mean of the droplet and sample areas (defined as $4\pi R_d R_c$ where R_d and R_c are the radii of the droplet and sample, respectively, idealized as surfaces of spheres). According to Boyer and Knipling (1965), the vapor transfer is given by

$$\frac{dm}{dt} = -\frac{La_m a_l}{La_m r_l + a_l r_a} \cdot \frac{c_o \bar{V}_w}{RT} (\Psi_o - \Psi_w), \tag{3.10}$$

where dm/dt is the rate of vapor transfer from the droplet ($g \cdot sec^{-1}$) and (Ψ_o - Ψ_w) is the driving force (MPa) with Ψ_o the water potential of the droplet and Ψ_w the water potential of the tissue. The term $c_o \bar{V}_w$ /RT is a constant at a particular temperature and serves to convert water potentials into the equivalent concentrations of water vapor (c_o is the saturation vapor concentration at the temperature of the chamber in units of $g \cdot m^{-3}$, R is the gas constant of 8.3143×10^{-6} $MPa \cdot m^{-3} \cdot mol^{-1} \cdot K^{-1}$, $\bar{V}_w$ is the partial molal volume in units of $m^3 \cdot mol^{-1}$, and T is the Kelvin temperature), and L is a constant that corrects droplet vapor concentration for the droplet temperature (has a value of 0.32 and indicates that water on the thermocouple is cool and thus at lower vapor pressure than it would be at chamber temperature).

When tissue is present, Eq. 3.10 indicates that the rate of transfer from the droplet to the tissue is determined by $La_m a_l/(La_m r_l + a_l r_a)$ but when a calibrating solution is present, r_l is zero and this relation becomes La_m/r_a. The lack of r_l causes diffusion to be faster for the calibrating solution than for the tissue even when the two have the same water potential. This difference in diffusion between the calibrating solution and the tissue is the source of the systematic diffusion error in calibrated psychrometers (Boyer and Knipling, 1965).

The size of the error is most easily estimated by noting that calibrated psychrometers determine the water potential by selecting the calibrating solution that gives the same thermocouple output as the sample being measured, that is by selecting for the same rate of vapor transfer for the calibrating solution and the sample, and reading the potential that gives this rate from the calibration curve. Accordingly, the water potential that gives this rate with tissue is Ψ_w and with calibrating solution is Ψ_a and these can be conveniently expressed as the ratio Ψ_w/Ψ_a. Because the rate of vapor transfer is equal for the calibrating solution and the tissue, Eq. 3.10 can indicate the equality as

$$\frac{La_m a_l}{La_m r_l + a_l r_a} (\Psi_w) = \frac{La_m}{r_a} (\Psi_a), \tag{3.11}$$

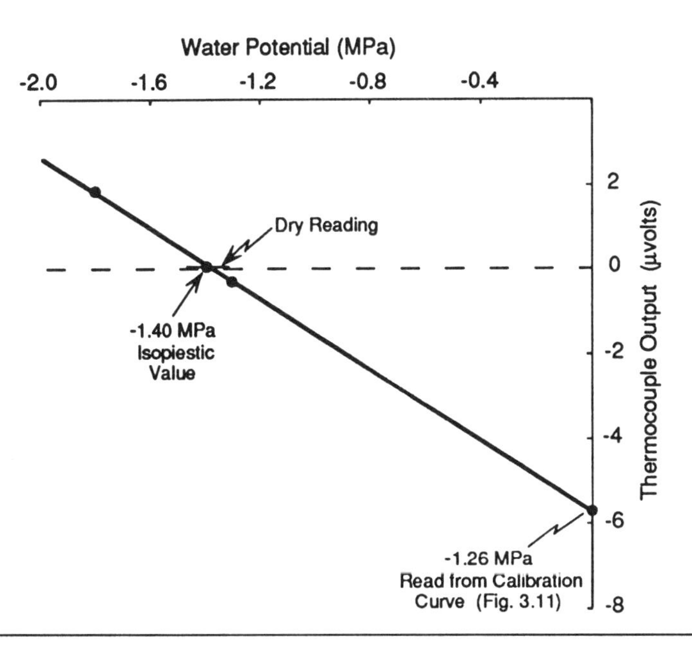

Figure 3.28. Diffusion error for a sample from a maize leaf. The true Ψ_w measured with the isopiestic psychrometer was -1.40 MPa while, with the same tissue sample, the same thermocouple measured only -1.26 MPa as a calibrated psychrometer (Richards and Ogata type) using the calibration curve in Fig. 3.11. Peltier psychrometers are subject to the same diffusion errors (Boyer and Knipling, 1965).

where the left side represents the tissue and the right side represents the calibrating solution, and $\Psi_o = 0$ because the droplet is water. Thus, $\Psi_w/\Psi_a = 1 + La_m r_l/a_l r_a$ indicating that Ψ_w is numerically larger than Ψ_a (more negative than Ψ_a). In other words, for the same rate of vapor transfer (and electrical output of the thermocouple), the true Ψ_w for the tissue is more negative than that determined from the calibration curve.

How much lower can be demonstrated in a psychrometer arranged as in Fig. 3.27. For this psychrometer having a 2-cm-diameter chamber and a 0.2-cm-diameter droplet, Eq. 3.11 gives a calculated $\Psi_w/\Psi_a = 1 + 0.032 r_l/r_a$, and for r_a of about 400 sec·m^{-1} for the air and a frequently encountered r_l of 1000 to 2000 sec·m^{-1} for the tissue (Boyer and Knipling, 1965), the calculated Ψ_w/Ψ_a is 1.08 to 1.16. Figure *3.28* shows that the Ψ_w and Ψ_a in this psychrometer for a maize leaf sample were -1.40 MPa measured isopiestically (Ψ_w) and -1.26 MPa measured in the same sample

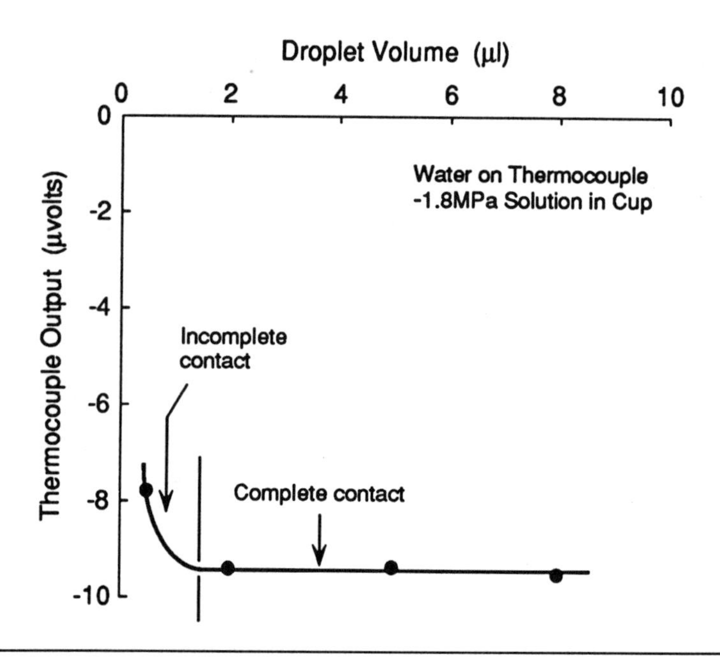

Figure 3.29. Effect of droplet size on thermocouple output. Measurements were made with an isopiestic thermocouple that completely contacted the droplet except when the droplet volume was decreased to 0.5 μl.

in the same psychrometer calibrated as a Richards/Ogata psychrometer (Ψ_a), that is, $\Psi_w/\Psi_a = 1.11$.

Therefore, in this example, the measured diffusion error in the calibrated psychrometer is 11% and is about the size of the calculated error (8-16%), which is significant. As can be seen from Eq. 3.11, the error varies with the size of r_l, and larger r_l causes more error. In contrast, the r_l does not affect an isopiestic measurement because $dm/dt = 0$ and the coefficient $La_m a_l/(La_m r_l + a_l r_a)$ has no effect. The vapor pressure of the droplet is the same as in the tissue, and $\Psi_0 = \Psi_w$ at the isopiestic point.

It should be noted that the size of the droplet on the thermocouple has little influence on the outcome. Figure 3.27 shows that the area of the droplet is much smaller than the area of the sample. The droplet approaches the behavior of a point source, and the rate of vapor transfer is not much affected by small changes in droplet size. Figure 3.29 shows that a measurement of *dm/dt* at various droplet sizes has no measurable effect as long as there is good contact between the thermocouple junction and the droplet.

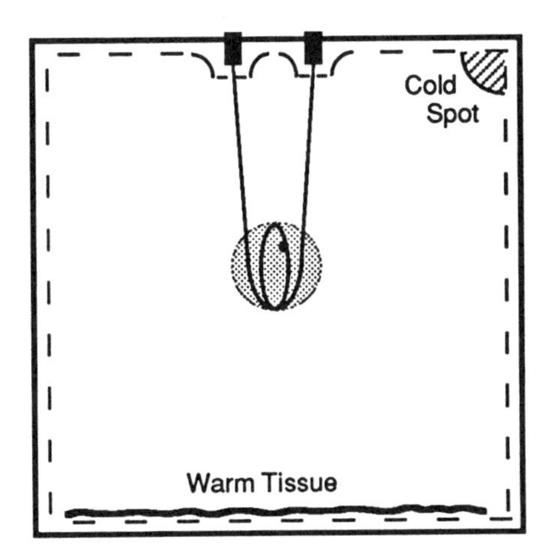

Figure 3.30. Effect of nonisothermal conditions (warm tissue or cold spots). External thermal gradients cause cold spots in the vapor chamber, resulting in condensation and disturbing the vapor conditions. Thermocouple readings cannot be corrected for these kinds of gradients. Internal gradients are caused by metabolic heat of the tissue. Measurements are corrected by noting the output of the thermocouple when dry and using the dry reading as the baseline for the output when wet.

ISOTHERMAL CONDITIONS

Uniform temperatures are important in psychrometry because the vapor pressure is temperature sensitive and the measurement circuit has its own thermoelectric activity. After uniformity is achieved, measurements can be made at different temperatures without a problem. Nonuniform temperatures can arise from two kinds of effects: temperature gradients imposed from the external environment and gradients generated internally, e.g., from the activity of cell metabolism inside the vapor chamber. Different techniques are used to correct the two types of gradients.

Externally imposed gradients cause one part of the vapor chamber to be colder than another as shown in Fig. 3.30 (cold spot). Water condenses from the chamber atmosphere, causing the water potential to appear very low to the thermocouple. A variation of ±0.1°C across the chamber results in a decrease in humidity equivalent to about 1.0 MPa. This cannot be corrected by measurements with the thermocouple because the cold area is remote and often small, and may alter the

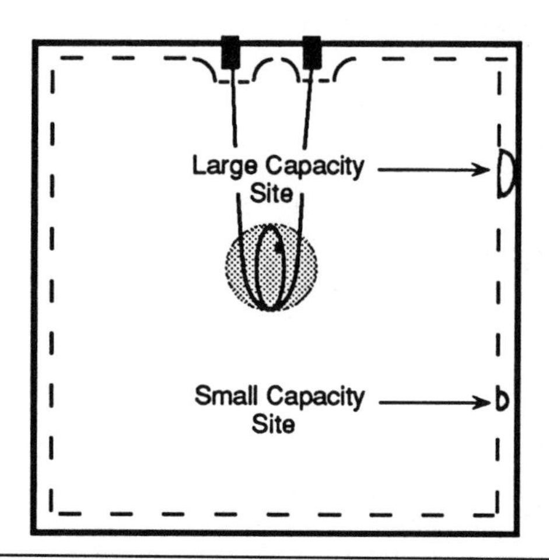

Figure 3.31. Sorption sites have large or small capacities in vapor chambers. Large sites take longer to equilibrate than small sites. Sites are present on walls and external tissue surfaces.

temperature of the thermocouple in unpredictable ways. Therefore, externally generated gradients must be avoided. To achieve an accuracy of ±0.01 MPa, temperatures must be uniform to ±0.001°C. This is usually achieved with a heat sink and insulation or with a water bath. Test for isothermal conditions in an isopiestic psychrometer by placing water on the spiral of an isopiestic thermocouple and on the bottom of the vapor chamber. If a steady reading of zero is obtained over long times, temperatures are uniform in the vapor chamber. In calibrated instruments, this test cannot be made.

Thermal gradients generated by metabolic heat are easier to deal with (Barrs, 1965). Metabolic heat affects the water potential readings by warming the thermocouple above the temperature it would normally have if the heat were absent (Fig. 3.30, warm tissue). As long as the heat does not raise thermocouple output more than 0.4 μV, measuring the output of the dry thermocouple can correct the output when the thermocouple is wet. In isopiestic measurements, the output of the dry thermocouple identifies the output at which a solution is isopiestic. Thus, isopiestic psychrometry automatically corrects for any heat produced by metabolism.

In addition to thermal effects in the vapor chamber, there also may be thermal effects in the measurement circuit. These can be detected by measuring the voltage produced by a dry thermocouple without any tissue or water in the vapor chamber. If a steady zero output is obtained over long periods of time, isothermal conditions are sufficient for the measurement circuit.

SORPTION EFFECTS

In the high humidity of a psychrometer chamber, any sites that can adsorb water vapor will rapidly remove water from the air. As the sites become hydrated, their water potential rises and the rate of sorption slows down. Eventually their water potential equilibrates with that in the psychrometer and the humidity begins to reflect that of the solution in the sample, allowing the sample water potential to be measured.

It is clearly advantageous to keep sorption to a minimum. All solid materials adsorb water vapor, but some are more sorptive than others. Rubbers and plastics have a high capacity to hold water adsorbed from the air (Brown and van Haveren, 1972). Metals, particularly when highly polished, have an intermediate sorptive capacity. Oils and waxes generally have a low capacity. However, impurities from manufacture can significantly increase the sorptive capacity of oils and waxes. So far, melted and resolidified Vaseline provides the least sorptive surface known to the author. All chamber surfaces should be coated with it except for the thermocouple detector and the sample.

Both the water potential of the sorptive sites and their capacity to sorb water affect the time needed for a thermocouple to reach the steady state, and the capacity is the most important (Fig. 3.31). If a large amount of water must be sorbed to raise the water potential, equilibration will be slow. If the capacity is small, equilibration will be rapid.

Sorption occurs on chamber walls and on plant samples. Plants generally are coated with waxes that have dust or other foreign material on the surface, and these are highly sorptive. The foreign material needs to be removed by washing as described earlier (Fig. 3.12). For plants that have relatively old leaves, e.g., conifer needles, some success has been obtained by soaking the intact needles in water for several hours. Always avoid dirty tissues or tissues having exposed cut surfaces or wounds because the surface can be more sorptive in these areas. For underground portions of the plant, most soil can be shaken away but some can be left attached if sampling is done with care because the adhering soil is in near equilibrium with the roots.

Although sorption delays the approach to steady conditions, the psychrometer responds rapidly to changes after sorption is completed (Boyer, 1969a; Boyer *et al.*, 1985). Isopiestic psychrometers can respond to a step change in vapor pressure immediately and have a time constant of about 30 to 45 sec.

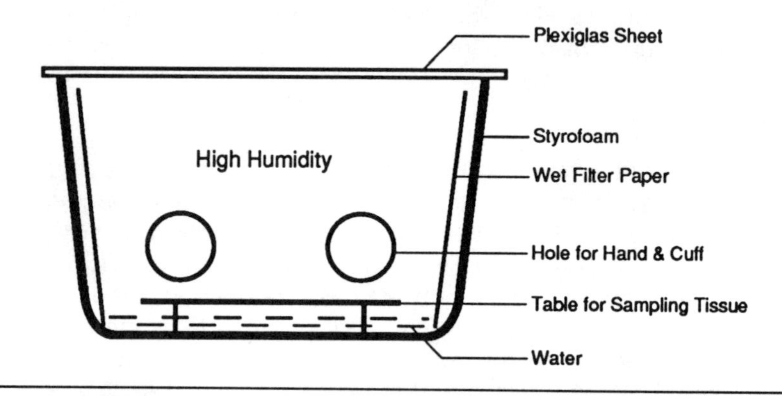

Figure 3.32. Glove box containing water-saturated air for sampling plant tissue and soils. The box has a clear top for viewing the sample on the table.

SAMPLING ERRORS

Evaporation after excision is probably the most frequent error in water potential measurements. To prevent it, tissue can be excised and loaded into the psychrometer rapidly (within 10 sec) or the tissue can be transferred immediately to a saturated atmosphere in a humid box where it can be loaded more slowly without significant evaporation.

A useful humid box can be constructed from a Styrofoam chest whose top has been replaced by a sheet of Plexiglas (Fig. 3.32). Cut two holes in the sides of the box for your hands. Attach plastic or rubber cuffs to the holes to seal around your wrists. Line the chest walls with wet filter paper and fill the bottom with water. Construct a small Plexiglas table to be placed inside. Place a dry paper towel on the Plexiglas table so that the sample is kept from contacting any wet surface. Insert your hands through the holes, sample the tissue on the table, load the Vaseline-coated psychrometer cup, and assemble the cup and psychrometer (Fig. 3.14).

Similar principles apply under field conditions. Carry the psychrometers to the field in a small Styrofoam chest and keep the chest out of direct sun. Complete the sampling rapidly or by using a glove box, as described earlier, then place the psychrometer units back into the Styrofoam chest. The sample should not be stored in the chest for longer than 15 to 30 min. In the laboratory, assemble the psychrometer system as soon as temperatures of the psychrometers approach laboratory temperatures.

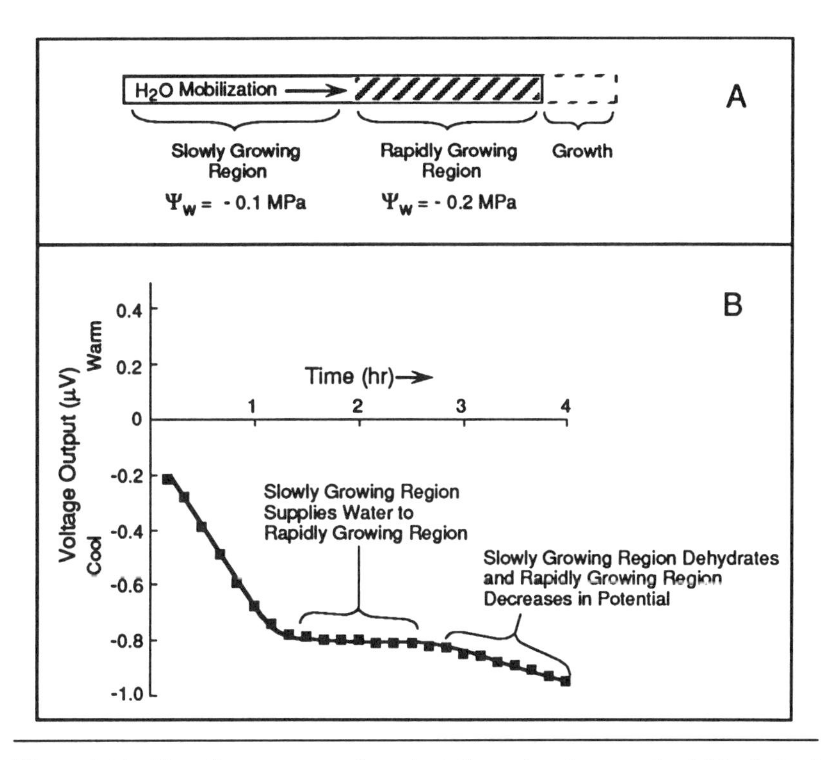

Figure 3.33. Growth can occur after plant tissue has been excised (A, dashed line). Water is mobilized from slowly growing to rapidly growing regions because the water potential is lower in the rapidly growing region. B) Psychrometer measurement in which rapidly growing and slowly growing soybean stem tissues were present in the vapor chamber. A stable output was obtained in the chamber between 1 and 3 hr after sampling, followed by a slow decline as water was withdrawn from the slowly growing region. The stable value gave the average water potential of the tissue and was the same as the potential in the intact plant. Provided readings were made between 1 and 3 hr, accurate measurements could be obtained. The slow decline that occurred after 3 hr did not relate to the intact plant and instead indicated changes occurring in the psychrometer chamber. Although the times vary, this is a typical behavior for all growing tissues.

GROWTH AFTER EXCISION

The water potential of tissue excised from growing regions may change slightly after excision (Boyer *et al.*, 1985). Excision disrupts the flow of water and solute into the enlarging tissue, but the cell walls continue to extend. The walls relax without water uptake until the turgor decreases to the turgor threshold where further wall growth ceases (Boyer *et al.*, 1985; Cosgrove *et al.*, 1984). The net effect is that the water potential becomes slightly lower (about 0.1 MPa) than it would have been in the intact plant (Boyer *et al.*, 1985). Figure 5.4 shows examples of this effect.

If mature tissue and growing tissue are excised together, water in the mature tissue is moved to the enlarging tissue. Wall relaxation is delayed and slow growth occurs (Fig. 3.33A). The water potential of the enlarging tissue continues to reflect that in the intact plant before excision (Boyer *et al.*, 1985; Matyssek *et al.*, 1988). The potential of the mature tissue gradually declines over a period of 2 to 3 hr as water is withdrawn to the enlarging tissue (Fig. 3.33B). When measuring the water status of enlarging tissue, it is often wise to include some attached, slowly growing or mature tissue if possible to give yourself time to complete the measurement rapidly while the tissue has a water potential similar to that in the intact plant.

VOLUME OF VAPOR CHAMBER

Because psychrometer measurements generally require 1 to 3 hr, an oxygen supply must be available for the tissue. Oxygen is supplied by the air enclosed with the tissue. Since most plant tissues have a low rate of oxygen consumption, small chambers are adequate. On the other hand, growing tissues may consume oxygen 10 times faster than mature tissues and larger chambers may be required. For a sample that is growing rapidly and has about 10 mg of dry weight, the chamber volume should be about 5 cm^3.

When tissue begins to run out of oxygen, the turgor decreases as membranes break down. The vapor pressure of cell water decreases and the tissue appears to become drier (not to be confused with the similar effects of wall relaxation in growing tissue). Eventually, the psychrometer indicates the osmotic potential for the protoplasts.

AVOIDING METABOLIC HEAT

The metabolic heat produced by plant tissue and by microorganisms in soil can be drained away by contact between the sample and the metal walls and bottom of the vapor chamber. Because these walls are coated with Vaseline, plant tissue can be pushed into the Vaseline to ensure good contact with the chamber surfaces (Fig. 3.20). Any residual heat is corrected by reading the output of the thermocouple when dry (Fig. 3.17). The correction is accurate only for small amounts of heat (dry reading to 0.4 μV). Never place more than one layer of tissue in the vapor chamber because heat is not readily drained from the inner layer, and the overheating will cause erroneous readings.

POTENTIAL GRADIENTS

Gradients in potential exist over large distances in both soils and plants. The gradients exist because water moving through the material flows through various frictional resistances, causing some parts to be drier than others. Also, gravity may cause gradients over long vertical distances. A knowledge of the size and location of the gradients is necessary for interpreting the measurements. Nonami *et al.* (1991) have shown that Ψ_w gradients exist between tissues in a leaf. For excised tissue, the potential is a volume average for that part of the gradient in the tissue at the time of sampling. For intact tissue, the potential appears to be that of the vascular supply that provides water to the sample (except in growing tissue, see earlier).

Leaves that are farthest from the water supply are usually drier than leaves closer to the water supply. Leaves that are most apically situated also tend to be more brightly illuminated than those at the base of the canopy, which further lowers their water potential. The reproducibility between measurements is enhanced when gradients are carefully considered during sampling. A particular problem occurs when making leaf gas exchange measurements in cuvettes because leaves outside the cuvette can have a different water potential from those inside. In this situation, gradients not only are present along the shoot but they are imposed by the experimental apparatus. Therefore, water potentials should be measured only with tissue inside the same cuvette used for the other measurements.

Appendix 3.1-Psychrometer Manufacturers

Wescor, Inc.
459 South Main Street
Logan, Utah 84321
(801) 752-6011

J.R.D. Merrill Specialty Equipment
R.F.D. Box 140A
Logan, Utah 84321
(801) 752-8403

Decagon Devices, Inc.
P.O. Box 835
NW 115 State Street
Pullman, WA 99163
(509) 332-2756

Isopiestics Company
2 Harborview Road
Lewes, DE 19958
(302) 645-4014

Appendix 3.2-Water Potentials (Osmotic Potentials) of Sucrose Solutions

Osmotic potentials for sucrose solutions are given in Table 3.1, calculated from the equations below. The calculations are according to Michel (1972). The osmotic potential (in MPa) of a sucrose solution is determined by

$$\Psi_s = -\phi m DRT/1000, \tag{3.12}$$

where ϕ, the osmotic coefficient, is a function of the molal concentration m of the solution, m is in mol·(kg $H_2O)^{-1}$, D is the density of water (g·m^{-3}) as a function of temperature T, R is the gas constant (8.3143×10^{-6} MPa·m^3·mol^{-1}·K^{-1}), and T is the Kelvin temperature (K). The osmotic coefficient ϕ for sucrose is generated by the empirical equation

$$\phi = 0.998 + 0.089m. \tag{3.13}$$

Substituting ϕ from Eq. 3.13 and $k = DRT/1000$ into Eq. 3.12, the osmotic potential is a function of the molality and the temperature-dependent coefficient k

$$\Psi_s = -(0.089\ m^2 + 0.998m)k. \tag{3.14}$$

Solving Eq. 3.14 for m:

$$m = -5.6067 + \sqrt{31.4355 - 11.236\ \Psi_s\ k^{-1}}\,. \tag{3.15}$$

TABLE 3.1. Water Potentials (Osmotic Potentials) of Sucrose Solutions. Molalities are in moles per kg of water, osmotic potentials are in bars, temperatures (T) are in Celsius.

T / Molality	12°	14°	16°	18°	20°	22°	24°	26°	28°	30°	32°	34°	36°
0.00	0.000	0.000	0.000	0.000	0.000	0.000	0.000	0.000	0.000	0.000	0.000	0.000	0.000
0.05	-1.187	-1.195	-1.203	-1.211	-1.219	-1.227	-1.234	-1.242	-1.249	-1.257	-1.264	-1.271	-1.279
0.10	-2.385	-2.401	-2.417	-2.433	-2.448	-2.464	-2.479	-2.495	-2.510	-2.525	-2.540	-2.554	-2.569
0.15	-3.593	-3.617	-3.641	-3.665	-3.689	-3.712	-3.736	-3.759	-3.781	-3.804	-3.826	-3.848	-3.870
0.20	-4.812	-4.844	-4.876	-4.908	-4.940	-4.971	-5.003	-5.033	-5.064	-5.094	-5.124	-5.154	-5.183
0.25	-6.041	-6.082	-6.122	-6.162	-6.202	-6.242	-6.281	-6.319	-6.358	-6.396	-6.433	-6.470	-6.507
0.30	-7.281	-7.330	-7.379	-7.427	-7.475	-7.523	-7.570	-7.616	-7.662	-7.708	-7.753	-7.798	-7.842
0.35	-8.531	-8.589	-8.646	-8.703	-8.759	-8.814	-8.870	-8.924	-8.978	-9.032	-9.085	-9.137	-9.189
0.40	-9.792	-9.858	-9.924	-9.989	-10.05	-10.12	-10.18	-10.24	-10.31	-10.37	-10.43	-10.49	-10.55
0.45	-11.06	-11.14	-11.21	-11.29	-11.36	-11.43	-11.50	-11.57	-11.64	-11.71	-11.78	-11.85	-11.92
0.50	-12.35	-12.43	-12.51	-12.59	-12.68	-12.76	-12.84	-12.91	-12.99	-13.07	-13.15	-13.22	-13.30
0.55	-13.64	-13.73	-13.82	-13.91	-14.00	-14.09	-14.18	-14.27	-14.35	-14.44	-14.52	-14.61	-14.69
0.60	-14.94	-15.04	-15.14	-15.24	-15.34	-15.44	-15.53	-15.63	-15.72	-15.82	-15.91	-16.00	-16.09
0.65	-16.26	-16.36	-16.47	-16.58	-16.69	-16.79	-16.90	-17.00	-17.11	-17.21	-17.31	-17.41	-17.51
0.70	-17.58	-17.70	-17.82	-17.93	-18.05	-18.16	-18.28	-18.39	-18.50	-18.61	-18.72	-18.83	-18.93
0.75	-18.91	-19.04	-19.17	-19.29	-19.42	-19.54	-19.66	-19.79	-19.91	-20.02	-20.14	-20.26	-20.37
0.80	-20.26	-20.40	-20.53	-20.67	-20.80	-20.93	-21.06	-21.19	-21.32	-21.45	-21.57	-21.70	-21.82
0.85	-21.61	-21.76	-21.91	-22.05	-22.19	-22.33	-22.47	-22.61	-22.75	-22.88	-23.02	-23.15	-23.28
0.90	-22.98	-23.14	-23.29	-23.44	-23.59	-23.74	-23.89	-24.04	-24.19	-24.33	-24.47	-24.61	-24.75
0.95	-24.36	-24.52	-24.69	-24.85	-25.01	-25.17	-25.32	-25.48	-25.63	-25.79	-25.94	-26.09	-26.24

TABLE 3.1 (continued). Water Potentials (Osmotic Potentials) of Sucrose Solutions. Molalities are in moles per kg of water, osmotic potentials are in bars, temperatures (T) are in Celsius.

T Molality	12°	14°	16°	18°	20°	22°	24°	26°	28°	30°	32°	34°	36°
1.00	-25.75	-25.92	-26.09	-26.26	-26.43	-26.60	-26.77	-26.93	-27.09	-27.26	-27.42	-27.57	-27.73
1.05	-27.14	-27.33	-27.51	-27.69	-27.87	-28.04	-28.22	-28.39	-28.57	-28.74	-28.90	-29.07	-29.24
1.10	-28.55	-28.74	-28.94	-29.13	-29.31	-29.50	-29.68	-29.87	-30.05	-30.23	-30.40	-30.58	-30.75
1.15	-29.97	-30.17	-30.37	-30.57	-30.77	-30.97	-31.16	-31.35	-31.54	-31.73	-31.92	-32.10	-32.28
1.20	-31.40	-31.61	-31.82	-32.03	-32.24	-32.44	-32.65	-32.85	-33.05	-33.24	-33.44	-33.63	-33.82
1.25	-32.84	-33.06	-33.28	-33.50	-33.72	-33.93	-34.14	-34.35	-34.56	-34.77	-34.97	-35.17	-35.37
1.30	-34.29	-34.52	-34.75	-34.98	-35.21	-35.43	-35.65	-35.87	-36.09	-36.30	-36.52	-36.73	-36.93
1.35	-35.75	-35.99	-36.23	-36.47	-36.71	-36.94	-37.17	-37.40	-37.63	-37.85	-38.07	-38.29	-38.51
1.40	-37.22	-37.48	-37.72	-37.97	-38.22	-38.46	-38.70	-38.94	-39.17	-39.41	-39.64	-39.87	-40.09
1.45	-38.71	-38.97	-39.23	-39.48	-39.74	-39.99	-40.24	-40.49	-40.73	-40.98	-41.22	-41.46	-41.69
1.50	-40.20	-40.47	-40.74	-41.01	-41.27	-41.53	-41.79	-42.05	-42.31	-42.56	-42.81	-43.05	-43.30
1.55	-41.70	-41.98	-42.26	-42.54	-42.81	-43.09	-43.36	-43.62	-43.89	-44.15	-44.41	-44.66	-44.92
1.60	-43.22	-43.51	-43.80	-44.08	-44.37	-44.65	-44.93	-45.21	-45.48	-45.75	-46.02	-46.29	-46.55
1.65	-44.74	-45.04	-45.34	-45.64	-45.93	-46.23	-46.52	-46.80	-47.09	-47.37	-47.64	-47.92	-48.19
1.70	-46.28	-46.59	-46.90	-47.21	-47.51	-47.81	-48.11	-48.41	-48.70	-48.99	-49.28	-49.56	-49.84
1.75	-47.82	-48.14	-48.46	-48.78	-49.10	-49.41	-49.72	-50.02	-50.33	-50.63	-50.92	-51.22	-51.51
1.80	-49.38	-49.71	-50.04	-50.37	-50.69	-51.02	-51.34	-51.65	-51.96	-52.27	-52.58	-52.88	-53.18
1.85	-50.94	-51.29	-51.63	-51.97	-52.30	-52.63	-52.96	-53.29	-53.61	-53.93	-54.25	-54.56	-54.87
1.90	-52.52	-52.88	-53.23	-53.58	-53.92	-54.26	-54.60	-54.94	-55.27	-55.60	-55.93	-56.25	-56.57
1.95	-54.11	-54.47	-54.84	-55.20	-55.55	-55.90	-56.25	-56.60	-56.94	-57.28	-57.62	-57.95	-58.28

TABLE 3.1 (continued). Water Potentials (Osmotic Potentials) of Sucrose Solutions. Molalities are in moles per kg of water, osmotic potentials are in bars, temperatures (T) are in Celsius.

T Molality	12°	14°	16°	18°	20°	22°	24°	26°	28°	30°	32°	34°	36°
2.00	-55.71	-56.08	-56.46	-56.83	-57.19	-57.56	-57.92	-58.27	-58.63	-58.98	-59.32	-59.66	-60.00
2.20	-62.20	-62.62	-63.04	-63.45	-63.86	-64.27	-64.67	-65.07	-65.46	-65.86	-66.24	-66.62	-66.70
2.40	-68.87	-69.34	-69.80	-70.25	-70.71	-71.16	-71.60	-72.04	-72.48	-72.91	-73.34	-73.76	-74.18
2.60	-75.71	-76.22	-76.72	-77.23	-77.73	-78.22	-78.71	-79.19	-79.67	-80.15	-80.62	-81.08	-81.54
2.80	-82.71	-83.27	-83.82	-84.37	-84.92	-85.46	-85.99	-86.52	-87.05	-87.56	-88.08	-88.59	-89.09
3.00	-89.88	-90.49	-91.09	-91.69	-92.28	-92.87	-93.45	-94.02	-94.59	-95.16	-95.72	-96.27	-96.81
3.20	-97.22	-97.88	-98.53	-99.18	-99.82	-100.5	-101.1	-101.7	-102.3	-102.9	-103.5	-104.1	-104.7
3.40	-104.7	-105.4	-106.1	-106.8	-107.5	-108.2	-108.9	-109.6	-110.2	-110.9	-111.5	-112.2	-112.8
3.60	-112.4	-113.2	-113.9	-114.7	-115.4	-116.1	-116.9	-117.6	-118.3	-119.0	-119.7	-120.4	-121.1
3.80	-120.3	-121.1	-121.9	-122.7	-123.5	-124.3	-125.0	-125.8	-126.6	-127.3	-128.1	-128.8	-129.5
4.00	-128.3	-129.1	-130.0	-130.9	-131.7	-132.5	-133.4	-134.2	-135.0	-135.8	-136.6	-137.4	-138.2
4.20	-136.5	-137.4	-138.3	-139.2	-140.1	-141.0	-141.9	-142.7	-143.6	-144.5	-145.3	-146.2	-147.0
4.40	-144.8	-145.8	-146.8	-147.7	-148.7	-149.6	-150.6	-151.5	-152.4	-153.3	-154.2	-155.1	-156.0
4.60	-153.3	-154.4	-155.4	-156.4	-157.4	-158.4	-159.4	-160.4	-161.4	-162.3	-163.3	-164.2	-165.2
4.80	-162.0	-163.1	-164.2	-165.3	-166.3	-167.4	-168.5	-169.5	-170.5	-171.5	-172.5	-173.5	-174.5
5.00	-170.9	-172.0	-173.2	-174.3	-175.4	-176.6	-177.6	-178.8	-179.8	-180.9	-182.0	-183.0	-184.1
5.20	-179.9	-181.1	-182.3	-183.5	-184.7	-185.9	-187.0	-188.2	-189.3	-190.5	-191.6	-192.7	-193.8
5.40	-189.1	-190.4	-191.7	-192.9	-194.2	-195.4	-196.6	-197.8	-199.0	-200.2	-201.4	-202.5	-203.7

Chapter 4

Pressure Probe

The pressure probe is the only instrument that can measure the water status of single cells. It consists of a transducer that monitors the pressure in an oil-filled microcapillary whose microscopic tip is inserted into a cell (Fig. 4.1A). The pressure necessary to prevent cytoplasm from entering the microcapillary equals the turgor pressure of the cell.

The probe can be used with isolated cells and with cells in a tissue. It gives data not only on the pressure but also on the elasticity of the wall, the hydraulic conductivity of the plasmalemma, and the selectivity of the plasmalemma for solute and water. The microcapillary can be used to collect a small sample of cell solution for later analysis or to inject solutions into cells.

The probe was initially developed for large cells of algae such as *Valonia* or *Nitella* (Steudle and Zimmermann, 1971) and it had large dimensions to allow various characteristics of the cells to be explored. A miniaturized version subsequently was developed for small cells using the principles of the larger instrument and is used for cells in tissues of multicellular plants (Hüsken *et al.*, 1978). The probe is particularly suited for detailed studies of the pressure-related properties of cells. For measuring the overall water status of the tissue, the psychrometer and pressure chamber require fewer measurements and are more suitable (see Chaps. 2 and 3). There is no commercial manufacturer of pressure probes. To help you make your own, some detailed drawings and addresses for equipment suppliers are given in Appendix 4.1.

Principles of the Method

Plant cells generally contain cytoplasm with a solute concentration higher than in the external solution, and water enters osmotically and stretches the wall. The resistance to stretching causes the wall to press on the cell contents and to establish a positive

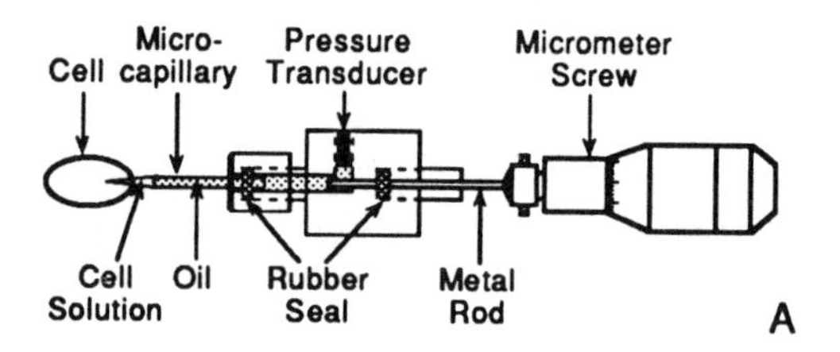

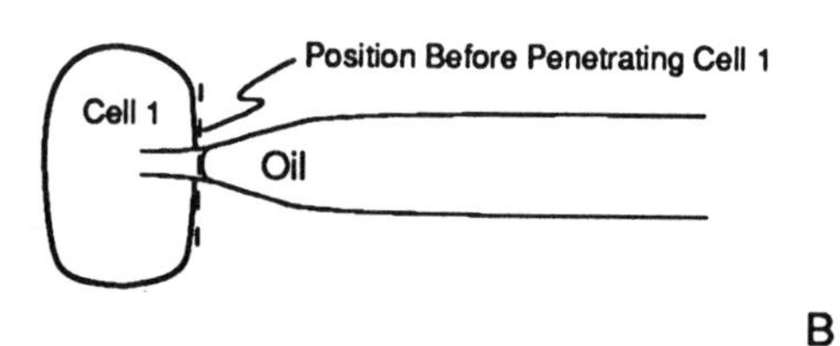

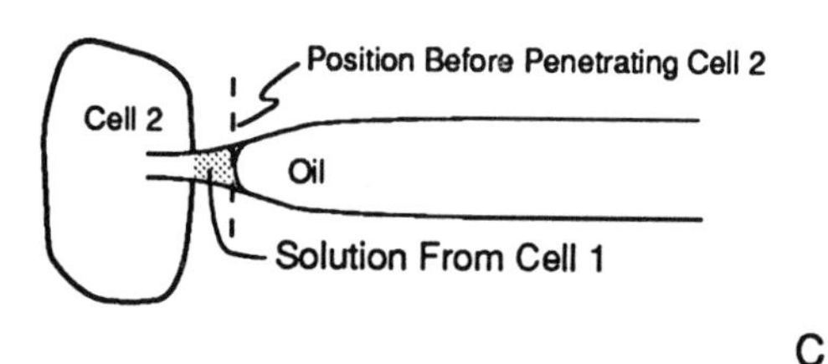

Figure 4.1. Diagram of a pressure probe (A) and position of cell solution during the measurement of turgor pressure (B and C). A) Pressure probe showing microcapillary, pressure transducer, metal rod, and oil-filled spaces (shaded). After inserting the microcapillary tip in a cell, a meniscus forms between the cell solution and the oil in the microcapillary and is observed with a microscope. The position of the meniscus is controlled by moving the metal rod in or out which raises or lowers the pressure in the oil. Turgor is measured by returning the meniscus to its position before entering the cell. B) Turgor measurement when the microcapillary is filled only with oil before entering cell. The meniscus is returned to the cell surface. C) Turgor measurement when the microcapillary already contains some solution from cell 1 (shaded) before entering cell 2. The meniscus is returned to its position before entering cell 2.

pressure, termed the turgor pressure or sometimes simply the turgor.[1] If the wall is highly resistant to stretching, a small amount of water entry causes a high pressure to develop. If the wall is easily stretched, the same amount of water entry causes only a small pressure. The wall thus controls the volume and pressure relations of the cell.

In an osmometer consisting of a rigid membrane separating a solution from pure water, a similar pressure develops that at its maximum is termed the osmotic pressure. However, the effect of solutes on water is most usefully expressed as an osmotic potential having pressure units. The pressure units indicate the maximum pressure that can be generated from the osmotic potential when a membrane is present but the osmotic potential is a fundamental property of the solution indicating the concentration, or more precisely, the activity of water in the solution. The osmotic potential exists whether or not a pressure is present.

It should be noted that pressures in cells are measured relative to the atmospheric pressure. They are thus differential or "gauge" pressures, and often are not at the maximum. Turgor pressures can vary from zero to pressures substantially above atmospheric, and tensions can develop that are below atmospheric. The developing pressures are determined to a large degree by membrane properties in the cell. In general, the membranes exclude solute and allow only water to pass through, creating a net water movement to the side with the lowest concentration of water (highest concentration of solute). If a membrane does not exclude solute, both solute and water move and there is essentially no net flow of molecules across the membrane. This lack of movement occurs despite the concentrated solution on one side of the membrane and pure water on the other. No pressure builds up. The membrane conductivity for water but selectivity against solute are thus the keys to the build up of cell pressure and volume generally.

---

[1]Pfeffer (1900) originally defined the turgor to be the stretch-induced strain in cell walls, and the turgor pressure to be the resulting pressure. Turgor pressure continues to refer to the pressure but recent usage of the term turgor has broadened to include either wall strain (turgidity) or the resulting pressure.

The pressure probe measures the pressure in the cell by puncturing the cell wall and plasmalemma with a sharpened tip of a microcapillary and measuring the pressure inside the microcapillary (Fig. 4.1A). When the tip enters the cell, the pressure in the cell pushes cytoplasm into the oil-filled microcapillary. The cytoplasm is pushed back into the cell with a metal rod that is forced into the oil, raising its pressure. As the pressure in the microcapillary rises, the cytoplasm is completely returned to the cell and the cytoplasm/oil meniscus is at its original position before puncturing. The pressure in the microcapillary is then the same as the pressure in the cell and is indicated by a pressure transducer.

The method relies on pushing the cytoplasmic solution back into the cell because the entry of a significant volume of cell solution into the microcapillary changes the volume and pressure in the cell, and an accurate turgor pressure for an intact cell can be determined only when there is no volume change. For a microcapillary completely filled with oil before puncturing, the meniscus is returned to the cell surface (Fig. 4.1B). However, if solution from previously punctured cells is already in the microcapillary, the meniscus is returned only to its original position before puncture (Fig. 4.1C).

Note that the pressure in the cell is measured by creating an opposing pressure in the oil of the microcapillary. The opposing pressure is the measured variable. Therefore, the measurement should be made at equilibrium where there is no significant flow in or out of the microcapillary or the tip should be large enough to allow flow without a significant resistance to ensure that the pressures are the same in the cell and the microcapillary.

Pressure Probe Theory

Figure 4.1A shows that the volume of the oil-filled compartment is small in the pressure probe to keep volume changes small and to minimize effects of temperature variation and apparatus elasticity on the pressure measurement. Designs with remote controls needing large oil volumes or large pressure transducers should be avoided. Rubber seals prevent leakage but allow movement of a metal rod and replacement of the microcapillary. The metal rod is screw driven and moves inward to increase the pressure and outward to

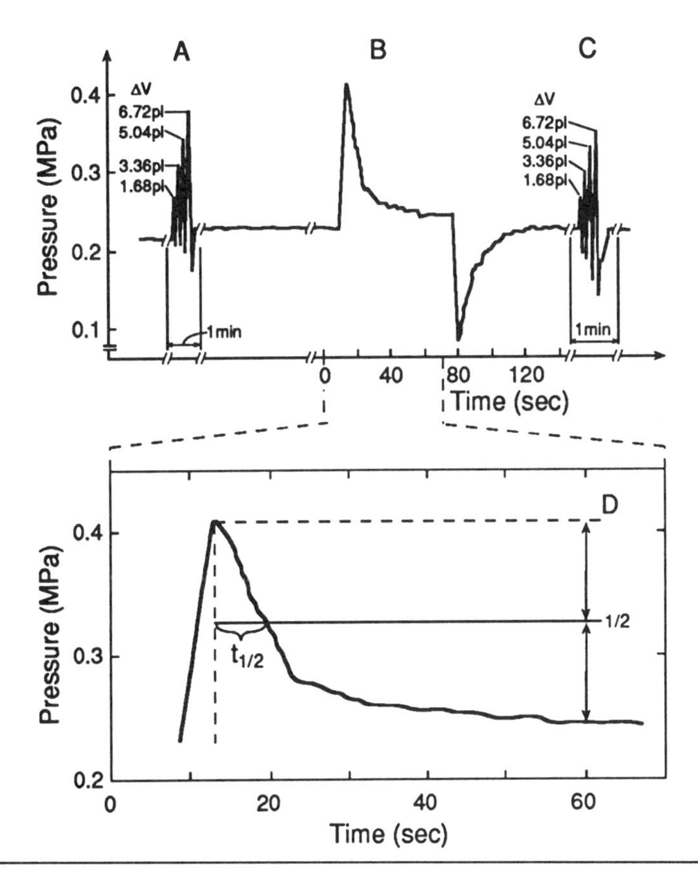

Figure 4.2. Turgor pressure when cell solution is injected into or removed from a *Tradescantia* subsidiary cell (next to guard cell in epidermis). A) Rapid injection and removal of cell solution to measure the turgor change for a known volume change ($d\Psi_{p(p)}/dV$). B) Rapid injection of cell solution and return of turgor toward original value as water flows out of the cell. Also shown is the reverse when cell solution is rapidly removed and water reenters the cell. C) Repeat of A. D) Expanded view of B showing the $t_{1/2}$ for water flow out of the cell. The $t_{1/2}$ is the time for half of the turgor to be lost and is used in calculating the hydraulic conductivity of the plasmalemma/cell wall (see Eq. 4.9). Adapted from Zimmermann *et al.* (1980).

decrease the pressure on the oil. The position of the oil/solution meniscus is observed under a stereomicroscope at a magnification of around 80X.

One of the most interesting aspects of osmosis is that water moves hydraulically through membranes and water diffusion plays only a negligible role. As a consequence, raising or lowering the cell pressure with a pressure probe causes an immediate hydraulically driven water flow out of or into the cell, which allows the probe to be used not only to measure pressure but also the nature of these flows. Figure 4.2B shows the immediate response of cell turgor pressure to pressure changes caused by the probe and the relaxation back toward the original pressure as water flows out of or into the cell in response (Zimmermann *et al.*, 1980). Such a behavior indicates that flow is hydraulically driven rather than diffusion driven because changing the pressure immediately changes the flow without altering the concentration of water. For this reason it is difficult to study water flow with isotopes of water or other molecular tracers that move by diffusion according to concentration differences. Hydraulically driven flow is also observed with pressure chambers.

As in Chap. 2, we will consider cells to have two compartments, the cell interior or protoplast and the cell exterior or apoplast (Fig. 2.3). In each compartment, particular components act on water and, for the protoplast, solute and pressure components are the main ones:

$$\Psi_{w(p)} = \Psi_{s(p)} + \Psi_{p(p)}. \tag{4.1}$$

For the apoplast, cell wall matrix and solute components are the main ones (the external pressure is atmospheric):

$$\Psi_{w(a)} = \Psi_{s(a)} + \Psi_{m(a)}. \tag{4.2}$$

The matric potential is mostly a negative pressure (tension) in the capillaries of the wall and associated vascular supply (xylem, see Fig. 2.3).

Under many conditions, the water potential is so similar on the inside and outside of the cell that they may be considered to be essentially equal:

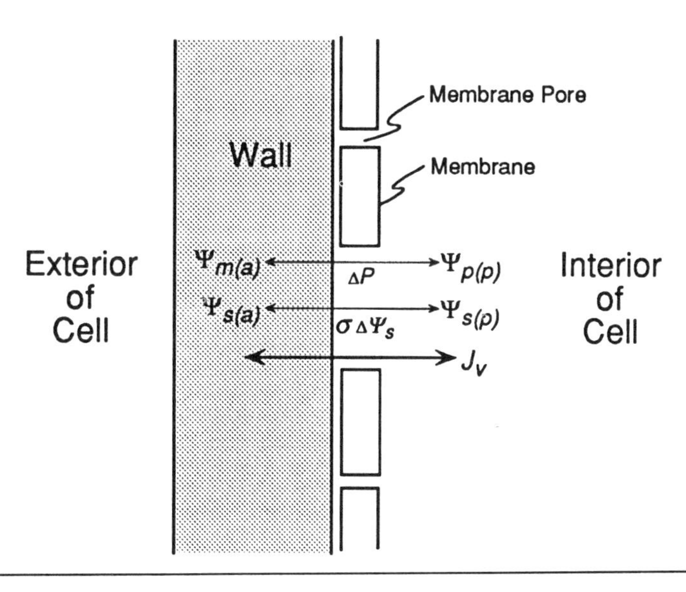

Figure 4.3. Diagram of the forces across the plasmalemma leading to water flow in cells in a tissue. The difference in pressure ΔP is determined by the turgor pressure inside and the matric potential (usually a tension) on the outside. The difference in osmotic potential $\Delta\Psi_s$ is expressed across the membrane according to the membrane reflectivity for solute shown as the reflection coefficient σ which can vary from 0 to 1. A membrane reflecting all solute and allowing only water to pass has a $\sigma = 1$ and allows all $\Delta\Psi_s$ to express itself. A membrane reflecting no solute has a $\sigma = 0$ and none of $\Delta\Psi_s$ will express itself. A cell with the latter membrane does not take up water despite a large concentration difference across the membrane. A net flow of molecules Jv results from $\sigma\Delta\Psi_s$ and ΔP.

$$\Psi_{w(a)} = \Psi_{w(p)} \tag{4.3}$$

and it becomes clear that the water potential inside the cell is balanced outside by any solute effects and matric forces.

The components outside differ according to where the cell is located. If a cell is in a tissue, solutes can be present outside (low $\Psi_{s(a)}$) together with a tension (low $\Psi_{m(a)}$) and the tensions extend to the water in the xylem and out into the soil. On the other hand, if a cell is not in a tissue and instead is directly surrounded by external solution, the wall is saturated with water and no tension exists (the pressure is atmospheric) and only $\Psi_{s(a)}$ contributes. Equation 4.3 then simplifies to

$$\Psi_{w(a)} = \Psi_{s(a)}. \qquad (4.4)$$

Caution needs to be used when interpreting changes in water potential or turgor measured with a pressure probe because of this distinction. The pressure probe is alone among the methods of this book in measuring the water status of the cell interior only. No information is given about conditions in the wall. Thus, when the turgor decreases inside, the cause cannot be attributed to external solutes unless the cell is surrounded by external solution at atmospheric pressure.

Water moves into and through the plant because of water potential differences or differences in a component of the water potential. When a potential difference exists across the plasmalemma, water moves at a rate determined by the conductivity of the plasmalemma and the size of the potential difference. Using the potentials of Eqs. 4.1 and 4.2 for the protoplast and apoplast (Fig. 4.3), the water movement can be described by the transport equation

$$J_v = Lp(\Psi_{m(a)} - \Psi_{p(p)} + \sigma(\Psi_{s(a)} - \Psi_{s(p)})), \qquad (4.5)$$

where J_v is the steady rate of volume movement (mostly water) across the membrane per unit membrane area $(dV/dt)(1/A)$ and has units of $m^3 \cdot m^{-2} \cdot sec^{-1}$, *Lp* is the hydraulic conductivity of the membrane $(m \cdot sec^{-1} \cdot MPa^{-1})$, $(\Psi_{m(a)} - \Psi_{p(p)})$ is the pressure difference across the membrane (the matric potential on the outside minus the turgor pressure on the inside of the membrane in MPa, see Fig. 4.3), $(\Psi_{s(a)} - \Psi_{s(p)})$ is the osmotic potential difference across the membrane (MPa, see Fig. 4.3), and σ is the reflection coefficient (dimensionless) indicating the fraction of solute prevented from crossing the membrane. The *Lp* represents the frictional effects encountered by water as it crosses the membrane, and a larger *Lp* shows that water crosses the membrane more rapidly for a given potential difference. The pressure probe can evaluate *Lp* by injecting or removing cell solution and observing the rate that water moves through the plasmalemma (and cell wall) in response.

For most cells, there is solute transport as well as water movement across the plasmalemma. Active metabolism usually is required for the solute but not the water. The solute also moves passively through the membrane. Both active and passive movements usually are slow compared to water, and the net movement of solute is largely independent of the movement of water. Therefore, the plasmalemma can be considered to be an ideal differentially permeable membrane with a reflection coefficient of essentially 1 for the solutes normally present inside a cell, and the hydraulic conductivity can be considered to apply only to water. Under these conditions, Eq. 4.5 reduces simply to

$$J_v = Lp(\Delta\Psi_w) \tag{4.6}$$

and water is driven across the plasmalemma/cell wall by the water potential difference ($\Delta\Psi_w$) between the two sides.

In special situations, this simplification may not hold. For small lipophilic solutes such as ethanol or isopropanol, σ is less than 1. Other solutes can alter membrane properties and cause σ to be less than 1 in which case internal solute may leak out. Cells that are suddenly subjected to high concentrations of solutes may shrink enough to cause the plasmalemma to separate from the cell wall (plasmolysis) and disrupt the plasmodesmata. Passive movement of solute becomes rapid and can be influenced by the rate of water movement. As a consequence, the exposure of cells to external osmotica can sometimes lead to these artifacts, and great care must be exercised. In these situations, it cannot be assumed that $\sigma = 1$ and the pressure probe, by measuring membrane properties, can detect when σ is less than 1.

SIGNIFICANCE OF REFLECTION COEFFICIENTS

The reflection coefficient is not a permeability coefficient but rather determines how much of the osmotic potential is harnessed in water transport. When σ is less than 1, the osmotic potential is similarly less than fully effective. If σ is less than 1, solute moves across the membrane but its rate is determined by a permeability coefficient for the solute and by any drag exerted by solute moving out of the cell while water moves in.

The osmotic potential can be measured inside and outside of cells but it does not indicate the reflection coefficient because the coefficient is a membrane property. There is no way to determine how much of the measured potential is contributing to flow without measuring the properties of the membrane. Depending on the reflection coefficient of the membrane, the osmotic effect of the solute can vary dramatically. Moreover, because the reflection coefficient describes a condition of the membrane, its effects are always present and cannot be avoided by making rapid measurements or allowing only small water flows. For this reason, osmotica generally do not simulate the natural dehydration of cells and are rarely used for measurements of cell water status.

WALL ELASTICITY

Sometimes it is useful to know how much water is required to change the cell turgor or water potential by a particular amount. The turgor pressure of the cell arises from the elasticity of the wall. The elasticity can be measured in terms of the bulk modulus of elasticity (ε in MPa, Chap. 2) according to

$$d\Psi_{p(p)} = \varepsilon \frac{dV}{V}, \tag{4.7}$$

which relates a change in internal pressure to the fractional change in water content (dV/V) of the cell. A larger ε indicates that a larger change in turgor will occur when the water content of the cell changes. In other words, a larger ε indicates that the wall is less elastic. It can readily be imagined that a rigid container (large ε) will experience a large pressure increase when the volume increases inside.

Similarly, the effect of dV/V on the water potential of the cell is described by the expression

$$\frac{dV}{d\Psi_{w(p)}} = \frac{V}{\varepsilon - \Psi_{s(p)}} = C, \tag{4.8}$$

which has been called the capacitance C of the cell (see Kramer and Boyer, 1995; Molz and Ferrier, 1982; Steudle, 1989 for derivations of Eq. 4.8). By injecting or removing cell solution rapidly with the pressure probe, it is possible to change the turgor and observe the volume change thus determining ε and C.

CELL KINETICS

Because the membrane properties in Eqs. 4.5 and 4.6 affect the rate of water flow, but the volume of water lost or more precisely the capacitance in Eq. 4.8 determines how fast the cell swells or shrinks, the effects of flow on swelling and shrinking provide a way to determine the hydraulic conductivity of the plasmalemma/cell wall complex and can be found by substituting Eq. 4.8 into Eq. 4.6

$$t_{1/2} = \frac{0.693V}{LpA(\varepsilon - \Psi_s)} = 0.693rC \qquad (4.9)$$

to give an expression describing how fast a cell shrinks or swells. In this equation, $t_{1/2}$ is the time for half the change in water potential or turgor (sec), A is the surface area of the cell (m^2), and r is the frictional resistance to water movement through the plasmalemma/cell wall ($1/LpA$). Kramer and Boyer (1995) give the derivation, which assumes the reflection coefficient equals 1.

Equation 4.9 shows that the cell acts much like an electrical circuit with a resistance and capacitance in series. The resistance r is mostly determined by the plasmalemma/cell wall and controls how fast water enters the cell. The capacitance C (Eq. 4.8) is determined by the size of the cell interior, the elasticity of its wall, and the internal osmotic potential, and these control how fast the potential changes for a unit change in the volume of water. The rate at which the water potential or turgor changes is the product of the resistance and capacitance, and an increase in either resistance or capacitance makes the change slower ($t_{1/2}$ longer).

Because the pressure probe moves water through the plasmalemma and cell wall during the *Lp* determination, *Lp* probably includes at least part of the cell wall conductivity as well as the plasmalemma conductivity. The contribution of the cell wall to this conductivity is not expected to be large, but measurements are needed to test this assumption. For many species, *Lp* ranges between 10^{-6} and 10^{-8} $m{\cdot}sec^{-1}{\cdot}MPa^{-1}$ (Kramer and Boyer, 1995). The range of values suggests that the plasmalemma/cell wall can vary in conductivity.

Figure 4.4. Miniature pressure probe mounted on a Leitz micromanipulator.

Using the Probe

After mounting the pressure probe on a micromanipulator (Fig. 4.4), place it on a vibration-free table. A stereomicroscope also should be on the same table and equipped with an eyepiece reticle to allow the solution/oil meniscus to be seen and measurements of the microcapillary dimensions and position to be made. The output of the pressure probe should be attached to a strip chart recorder to provide a continuous record of the pressures. If available, a data logger can be used to store the data for later analysis. In some advanced systems, it has been possible to view the meniscus and acquire and analyze the data with a computer. With your electrical equipment operating, go through the following procedures to ensure that accurate measurements are achieved.

FILLING WITH OIL

The pressure probe should be filled with silicone oil of the least viscous grade (Wacker AS 4 or equivalent). The oil should be clean and may need to be filtered if it contains significant foreign particles.

Figure 4.5. Calibration system for the pressure transducer mounted on a miniature pressure probe. A gas line replaces the microcapillary. The probe is filled with silicone oil and gas pressure is applied to the meniscus. The gas pressure is measured with a test gauge.

Usually, a syringe having a fine needle is filled with the oil and the borings are filled in an order that drives out all air. Be careful not to expose the pressure transducer to large pressures or bump it with the needle. Mount the rod and transducer only partially, and fill the borings until oil comes out. Then screw the rod and transducer mountings into place tightly, being careful to have an opening present to release the pressure generated as these fittings are tightened. After filling with oil, inspect all parts of the probe to ensure that no air bubbles are present. If bubbles are observed, repeat the filling procedure.

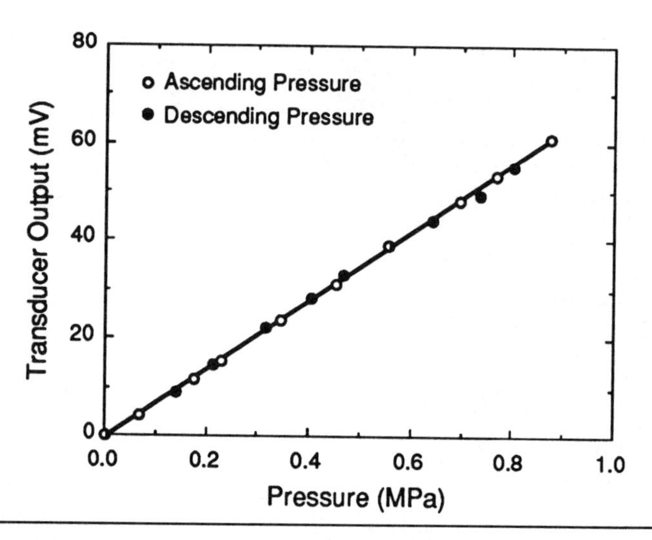

Figure 4.6. Calibration of the output of a pressure transducer with a test gauge shown in Fig. 4.5. From Nonami (1986).

CALIBRATION

The pressure transducer should be calibrated by placing a tube in the socket usually occupied by the microcapillary (Fig. 4.5). The open end of the tube is connected to a compressed gas cylinder. The pressure from the cylinder is increased and applied to the oil-filled probe. The pressure is measured with a high quality pressure gauge (test gauge or master test gauge, see Fig. 4.5) and the electrical output of the transducer is noted. After a series of pressures, a calibration curve can be constructed as shown in Fig. 4.6. The plot should be linear and, because the transducer should be the type to indicate gauge pressure, the output should be zero at atmospheric pressure.

MICROCAPILLARY

The microcapillary is made by heating a capillary (1 mm outside diameter, thin wall) at a point about halfway between the ends and rapidly pulling the ends apart. The melted part is drawn to a fine point that seals, and the two halves of the capillary are formed into near identical microcapillaries.

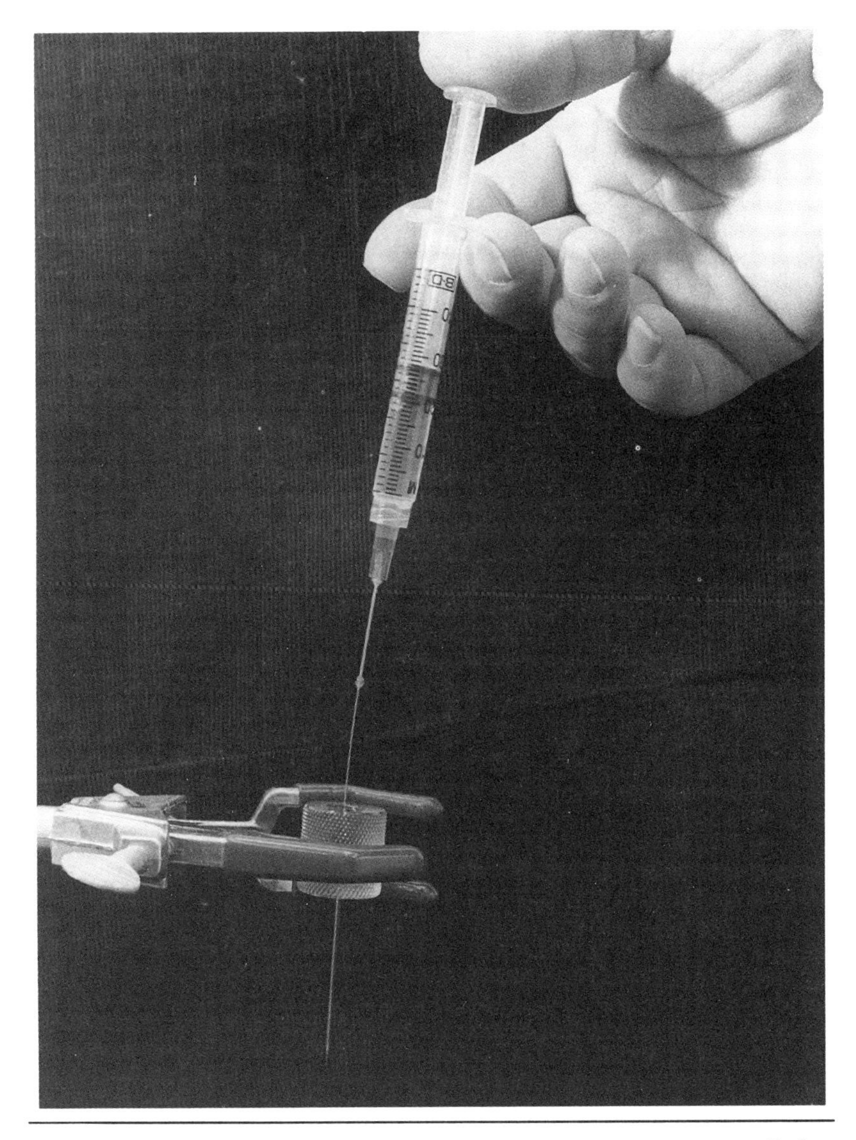

Figure 4.7. Filling a microcapillary with silicone oil. The syringe needle has been lengthened by inserting a fine tube into the needle and sealing with a drop of epoxy.

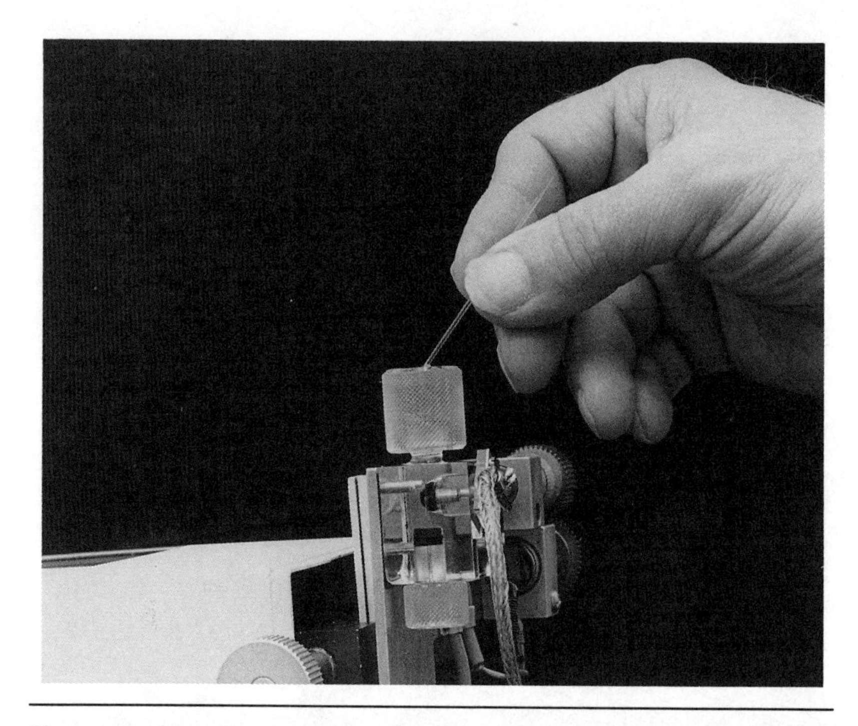

Figure 4.8. Mounting a microcapillary on a miniature pressure probe. The probe has been filled with silicone oil sufficiently to flood the surface of the holder, and the microcapillary has been filled with oil. Holding the open end of the microcapillary under the oil on the flooded surface excludes all air bubbles. The microcapillary is then placed upright in the holder, mounted in the seal, and the holder is tightened while watching the output of the pressure transducer on a strip chart recorder. Be careful not to overpressurize the oil when sealing the microcapillary in place.

The rapid pull of the capillary is most reproducible when a vertical electrode puller such as that made by Kopf (Model 720) is used to heat and do the pulling. Horizontal pullers have the disadvantage that the microcapillary tip sags under the force of gravity during the pulling, which bends the tip. A vertical puller has the advantage that the pull of gravity is symmetrical and no sagging occurs.

FILLING THE MICROCAPILLARY WITH OIL

Mount a microcapillary in the seal of the pressure probe (Fig. 4.7) for filling with oil. A long thin syringe needle can be constructed by gluing a thin metal tube (such as a thin-walled stainless-steel tube part number Q-HTX-27TW, Small Parts Company, Miami Lakes, Florida) into a syringe needle. Attach the needle to a syringe filled with silicone oil and insert the thin tube into the microcapillary. Move oil into the microcapillary while slowly withdrawing the tube until the microcapillary is completely filled. Place the filled microcapillary on the pressure probe and seal without admitting air bubbles (Fig. 4.8).

LEAKS

Test the sealed pressure probe for leaks by screwing the rod into the oil until the output of the pressure transducer becomes high (up to 1 MPa). After high pressure is established, observe the pressure for several hours. Only a slow decline in pressure should be observed. If the pressure decreases rapidly, double check that the tip of the microcapillary is still sealed (observe that no oil is leaking using a stereomicroscope). Provided the tip seal is good, the leak is most likely to be in the seal for the microcapillary, transducer, or rod. These seals are rubber gaskets that degrade after extended use and need to be replaced. If excessive leaks are observed, replace the seals and repeat the leak test. Seals remain good for 2-3 years.

BEVELING THE MICROCAPILLARY TIP

The tip of the microcapillary needs to be opened to allow the silicone oil to contact the cytoplasm inside the cell. Although the tip can be forced against a surface and broken open, the jagged edge reduces the rate of success in measuring cell pressures. Typically, we obtain success rates of 20-30% with tips opened in this fashion. Success rates of 80-90% are achieved when the tip is opened by beveling.

For beveling, the tip is brought against a rotating disk coated with diamond dust (such as that made by Narishige Model EG-4) while the silicone oil is under pressure. The tip is angled at about 45° to the surface. When an oil streak appears on the rotating disk, the tip has been beveled sufficiently to open and is usually ready to insert into a cell. Tip openings are about 1 µm. For some cells such as the large internodal cells of *Chara*, the tips tend to plug and beveling at a larger angle and for a longer time reduces plugging. Tip openings of 5 µm or more may be required (Zhu and Boyer, 1992), and the large diameter

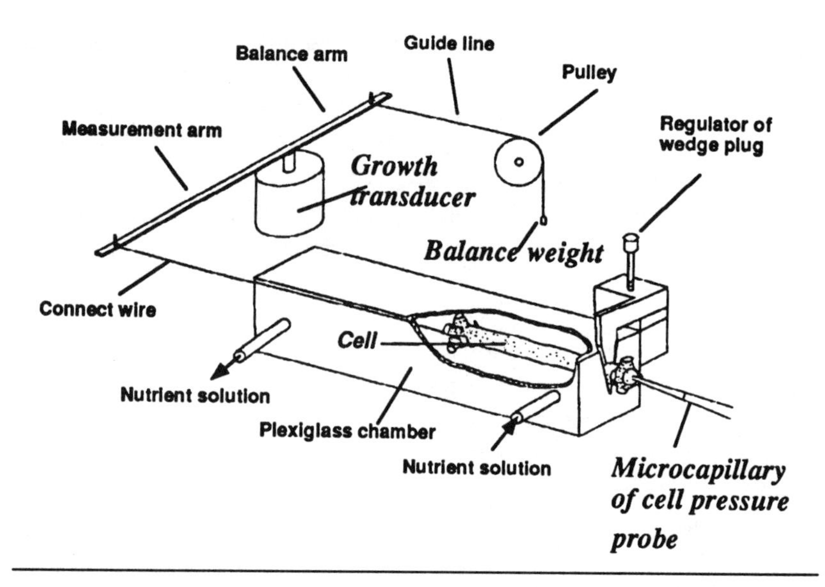

Figure 4.9. Support for a *Chara* cell. The microcapillary tip is inserted in the rigidly mounted end of the cell, and nutrient solution fills the chamber. In this setup, a position transducer is attached to the other end of the cell. From Zhu and Boyer (1992).

apparently prevents cytoplasm from flowing over the opening and sealing it.

MICROMANIPULATOR

The pressure probe must be mounted on a micromanipulator in order to allow small controlled movements of the microcapillary. A micromanipulator such as the one built by Leitz (Fig. 4.4) has a massive base and a large moving platform that are ideal for minimizing vibration and mounting the probe with its accompanying instrumentation.

CELL SIZE

The minimum cell diameter is 25 µm for routinely measuring turgor pressure, elastic modulus, and hydraulic conductivity with the probe. Occasionally, turgor can be measured with smaller cells (as small as 15 µm diameter). Cell properties other than turgor are difficult

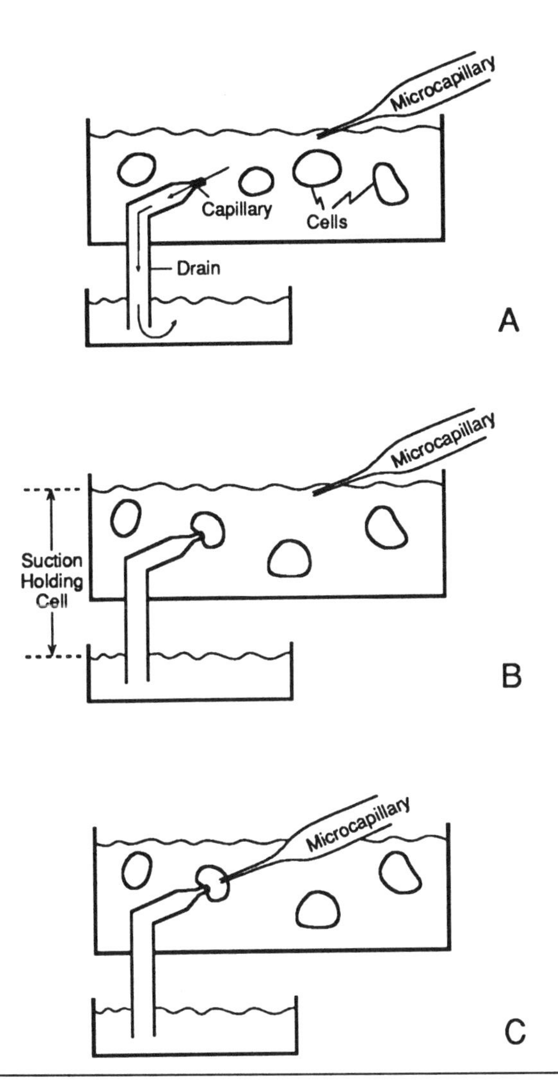

Figure 4.10. Support for a single small cell. Suction is created around the opening of a microcapillary with a fire-polished edge. A cell is swept against the opening by the suction and held there. The pressure probe is advanced into the front of the cell while the microcapillary in the back provides support. The suction can be varied by adjusting the height of the drain tank. The dimensions are not to scale.

to measure because there is more cell disturbance. Apparently, the limiting cell diameter is fixed by the diameter of the microcapillary tip because openings of 1-2 μm in the tips cause relatively large wounds in small cells.

How to Make Measurements

TURGOR PRESSURE

The cell or tissue is typically supported on the side opposite the one for insertion of the microcapillary tip. Mounts for cells or excised tissues may consist of culture vessels with the end of the specimen held immobile for the probe but the rest covered by nutrient solution as in Fig. 4.9. Surrounding the specimen with a nutrient solution is essential to minimize evaporation, and great care must be taken to prevent evaporation from altering the turgor of the measured cells (see Precautions). For single cells, it may not be possible to clamp one end because of the small size of the cell. In this situation, another microcapillary is constructed and the tip is ground until the opening is several micrometers in diameter. The tip is lightly fire polished to provide smooth rounded surfaces at the opening. The tip is inserted into the medium containing the cells and the medium is sucked into the microcapillary by forming a drain with a tube attached to the other end of the microcapillary. Under the microscope, the tip is moved among the cells in the solution until a cell is sucked against the tip, blocking it, and remaining in place because of the suction. The suction should be only strong enough to hold the cell in place without distortion. One design shown in Fig. 4.10 immobilizes the cell from one side and allows the microcapillary of the pressure probe to be pushed from the other side. Another approach is to catch a large number of cells on a screen using slight suction and use the pressure probe on individuals that have been immobilized.

Intact plants may be supported by placing a rigid bar behind the probed region as in Fig. 4.11. A saturated atmosphere can be used to minimize evaporation. Cover the surfaces of the chamber or work room with water and allow the water to vaporize. Saturation of the air can be detected by weighing wet filter paper. When no weight loss is observed, the atmosphere is saturated. However, because part of the

Figure 4.11. Support for a soybean stem. A rigid bar is placed behind the stem to support the tissue into which the microcapillary will be driven. Note the coating of petrolatum (Vaseline) on the surface of the seedling to prevent evaporation. From Nonami and Boyer (1993).

tissue must be illuminated, local heating from the light source can cause evaporation into a saturated environment (Nonami *et al.*, 1987). To avoid this, cover the plant part with a thin layer of petrolatum (Vaseline) and insert the microcapillary through the Vaseline into the tissue as in Fig. 4.11.

The need for illumination arises because the meniscus must be observed between the oil and the cytoplasmic solution. The specimen and microcapillary are illuminated at an angle that causes the meniscus to be seen as a bright surface in the microcapillary. A fiber optic cable provides a simple means of directing light to the microcapillary, and small light-emitting diodes also have been used successfully.

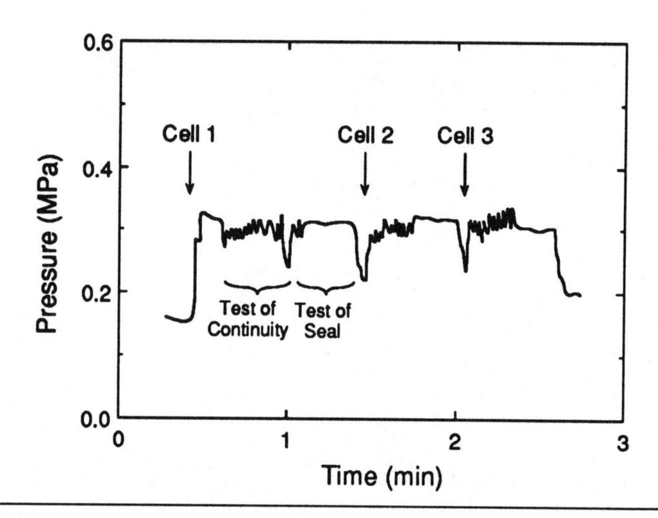

Figure 4.12. Typical recorder tracing of turgor pressure in three cells in a soybean stem. The liquid continuity between the cell and pressure transducer is tested by rapidly varying the pressure in the probe and observing the movement of the meniscus in and out. If the meniscus does not move, no liquid is moving and continuity does not exist. The seal with the cell is tested by observing the pressure for about 10 - 15 sec. If the pressure drops, a leak is present. From Nonami *et al.* (1987).

1) With the sample supported and protected from evaporation, the microcapillary tip and sample are brought into view under a stereomicroscope at low magnification. A small amount of silicone oil is usually being released from the tip of the beveled microcapillary. The magnification is increased to about 80X. Although the opening at the tip is too small to be seen at 80x, the tip should appear distinct with no fuzziness. If the tip is fuzzy, there is too much vibration. Determine whether the vibration is caused by table movement or air movement, and take steps to eliminate it (see Precautions).
2) Move the microcapillary slowly forward in a straight line and observe the silicone oil close to the tip as the microcapillary enters the cell or tissue. When a cell is penetrated, the cytoplasm will enter the microcapillary abruptly and a bright oil/cytoplasm meniscus should be seen to jump back into the microcapillary. This jump is diagnostic for

penetrating a cell. For a tissue, use the eyepiece reticle to note the depth that the tip has penetrated so that the location of the measured cell can be determined later.

3) Slowly bring the meniscus toward the cell by pushing the rod into the oil of the pressure probe. Stop the meniscus at the original position before entering the cell. If the probe was completely oil filled, this return pressure brings the meniscus to the cell or tissue surface. If the probe contained solution before entering the cell, the meniscus is returned only to the original position before penetrating the cell (Figs. 4.1B and 4.1C). This pressure is the turgor pressure of the cell.

4) With slight movements of the rod, move the meniscus short distances around the original position (Fig. 4.12). Small pressure fluctuations should move the meniscus freely, and the pressures should fluctuate around a mean that is the same as the turgor of the cell, which establishes that hydraulic continuity exists between the cell and the pressure transducer.

5) After testing hydraulic contact, observe the turgor for about 15 sec (Fig. 4.12). The turgor should remain stable, indicating that a good seal exists between the tip and the cell. If the turgor diminishes, there is a leak and the tip should be moved to the next cell repeating Steps 2-5.

6) There can be significant variation in the turgor pressure between cells in a tissue, and representative turgors can be determined only with a statistically relevant sample (an example is in the following chapter, see Fig. 5.2). For most tissues, about 7-10 cells provide a large enough sample to establish a mean and variance.

OSMOTIC POTENTIAL

After measuring the turgor pressure, the osmotic potential can be determined by removing a sample of the cell solution and measuring its osmotic potential in a freezing point osmometer.

1) With the microcapillary tip in the cell, note the position of the meniscus, move the meniscus back into the microcapillary, note the distance, and rapidly withdraw the tip from the cell. The microcapillary now contains a sample of solution from the cell. The time to move the meniscus back and withdraw the tip should be no

more than 1-2 sec in order to avoid diluting the sample with water entering the cell as the turgor decreases (Malone *et al.*, 1989).

2) Swiftly transfer the sample from the microcapillary to a sample well in a nanoliter freezing point osmometer (Clifton Technical Physics, Hartford, NY). By moving small distances, the sample can be transferred to the well within a few seconds. The rapid transfer is necessary to avoid evaporation of the sample from the tip of the microcapillary. The well should be filled with immersion oil of the most viscous grade (Cargille Type B or NVH) to immobilize the sample droplet in the well.

3) View the sample droplet under a compound microscope at 300-400X using an objective having a long focal length. It should be possible to observe volumes as small as 10 picoliters. Make certain that the droplet remains suspended in the oil and does not contact the side of the well. Contact distorts the droplet and alters the freezing point. Although it is possible to calibrate the freezing point for contact with the sides of the well (see Malone *et al.*, 1989; Malone and Tomos, 1992), the simplest approach is to prevent contact.

4) Freeze the droplet and begin the thaw cycle, noting the temperature at which the last ice crystal just disappears. The freezing point is defined as the temperature of disappearance of the last crystal because the presence of significant ice, which is relatively pure water, concentrates the cell solution and prevents it from being the original concentration. The cell solution is closest to its original concentration as the last ice disappears. This procedure should be repeated with the sample and should give reproducible freezing points.

5) Express the freezing point as the number of degrees Centigrade below zero (ΔT). Calculate the osmotic potential from the equation

$$\Psi_{s(p)} = \frac{-2.27\Delta T}{1.86}. \qquad (4.10)$$

The $\Psi_{s(p)}$ has units of MPa and is at 0°C. For $\Psi_{s(p)}$ at other temperatures, multiply the osmotic potential at 0°C by T/273 where T is in Kelvin.

6) If samples have been obtained from the first cell in contact with the microcapillary, there is no other solution in the microcapillary and the

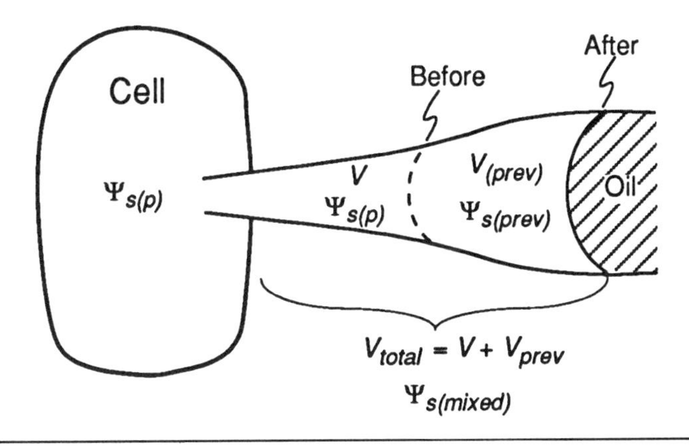

Figure 4.13. Sampling cell solution with a pressure probe when mixing occurs in the microcapillary. If a significant volume of solution $V_{(prev)}$ already exists in the microcapillary before entering a cell, removing a new volume V from the cell can mix with $V_{(prev)}$. By sampling cells around the test cell, the $\Psi_{s(prev)}$ can be measured. Then for the test cell, noting the position of the meniscus before sampling (Before) and after sampling (After) allows the previous volume ($V_{(prev)}$) and sample volume (V) to be calculated. Measuring $\Psi_{s(mixed)}$ from the mixed volume allows $\Psi_{s(p)}$ to be calculated for the test cell according to Eq. 4.11.

$\Psi_{s(p)}$ is calculated directly from Eq. 4.10. If the microcapillary already contains some solution from previously penetrated cells, the sample from the cell can be kept small so that it enters only the tip of the microcapillary where it has little chance to mix with other solution in the microcapillary. In this situation, $\Psi_{s(p)}$ also can be calculated by Eq. 4.10 without correction.

7) If the sampled solution moves into the main body of the microcapillary where other solution is present, there is a chance of mixing with solution already in the microcapillary (Fig. 4.13). Typically, complex tissues require filling the microcapillary with solution from surface cells in order to keep the meniscus in view as the tissue is penetrated more deeply. Determining the $\Psi_{s(p)}$ of these surface cells gives the osmotic potential of the solution already in the microcapillary. A correction for mixing can be made if the volume of this solution and the sample solution also are known (Fig. 4.13). After

measuring the osmotic potential of the cells close to the surface in the vicinity of the deeper cell to be sampled, determine the volume of the solution previously in the microcapillary and the volume from the sampled cell by calculating from the position of the meniscus noted in Step 1 before and after taking the sample into the microcapillary (Fig. 4.13). The volume of the interior of the microcapillary can be determined by approximating the tip as a cone or series of cones and the main axis as a cylinder. The volume of the sample (V) is then determined by subtracting the volume before sampling (V_{prev}) from the volume after sampling (V_{total}). The measured osmotic potential is corrected for mixing with the solution already in the microcapillary to give the original $\Psi_{s(p)}$ of the sampled cell using a relation similar to that in Eq. 3.8 but modified to apply to the microcapillary

$$\Psi_{s(p)} = \Psi_{s(mixed)} + \frac{V_{(prev)}}{V} (\Psi_{s(mixed)} - \Psi_{s(prev)}), \qquad (4.11)$$

where the subscripts *(mixed)* and *(prev)* indicate the mixed sample and the previous solution in the microcapillary, respectively. According to this calculation, the $\Psi_{s(mixed)}$ is between the true $\Psi_{s(p)}$ of the sampled cells and the $\Psi_{s(prev)}$ of the previously penetrated cells in proportion to the relative volumes of each in the microcapillary. For example, for $V_{(prev)} = 0.4V_{(total)}$ and $V = 0.6V_{(total)}$, and a measured $\Psi_{s(prev)}$ of -1.2 MPa and $\Psi_{s(mixed)}$ of -1.4 MPa, the calculated $\Psi_{s(p)}$ is -1.53 MPa. As you can see, the corrections are usually small (a few hundredths of an MPa) but they should be made for accurate work.

Note that it is incorrect to avoid this correction by expelling the solution in the microcapillary before entering the sampled cell. If the solution in the microcapillary is expelled, the meniscus becomes invisible and the microcapillary must be refilled with the solution from the sampled cell. The total sampled volume is thus large and can be larger than the volume of the sampled cell. The osmotic potential is markedly diluted by water entering the cell in this situation. Instead, short times (Malone *et al.*, 1989; Malone and Tomos, 1992) and small sample volumes corrected for mixing (Nonami and Boyer, 1993) should be used to accurately determine cell osmotic potentials with a pressure probe.

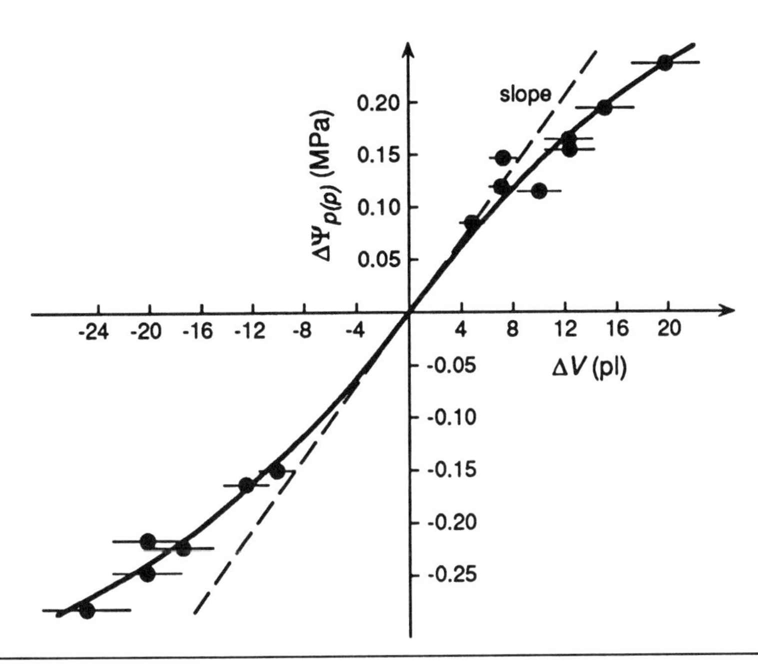

Figure 4.14. Change in turgor pressure with change in volume of cell solution measured as in Fig. 4.2A. In this example, small volumes ΔV are injected into the cell and cause positive $\Delta\Psi_{p(p)}$, and small volumes are removed causing negative $\Delta\Psi_{p(p)}$. The dashed line gives the slope which is $d\Psi_{p(p)}/dV$ for calculating the elastic modulus ε. The deviation from the dashed line is caused by finite amounts of water moving through the plasmalemma when the pressure change is large. From Tomos *et al.* (1981).

The microcapillary tip is usually in the central vacuole of the cell, and the measured osmotic potential is for the vacuole to the largest extent. However, the osmotic potential is essentially the same throughout the cytoplasm and vacuole (see Kramer and Boyer, 1995), and the results with the pressure probe sample should apply throughout the cell. Using various micromethods of analysis, the sample of cell solution may be used for other vacuolar determinations such as the elemental content (Malone *et al.*, 1989), metabolite levels (Harris and Outlaw, 1991; Outlaw and Lowry, 1977), or enzyme activities (Hite *et al.*, 1993).

ELASTIC MODULUS

The wall ε can be determined by changing the volume of the cell with the pressure probe while measuring the pressure, then measuring the cell volume microscopically.

1) Measure the turgor pressure as described earlier.
2) Note the position of the meniscus in the reticle of the eyepiece of the stereomicroscope and rapidly pull a small amount of solution from the cell into the microcapillary, note the new position of the meniscus and the new turgor pressure, and rapidly return the meniscus to its original position. This step should take no more that 1-2 sec to minimize water movement across the plasmalemma (see rapid pressure transients in Figs. 4.2A and 4.2C).
3) Repeat Step 2 with larger volumes of cell solution, working to decrease and increase the pressure as rapidly as possible. This allows $d\Psi_{p(p)}/dV$ to be calculated for decreasing $\Psi_{p(p)}$ (Fig. 4.14).
4) Reverse the procedure in Step 2 by rapidly injecting a volume of solution into the cell from the microcapillary and noting meniscus positions and pressures. Inject increasing volumes of solution as in 3). This allows $d\Psi_{p(p)}/dV$ to be calculated for increasing $\Psi_{p(p)}$ (Fig. 4.14).
5) Measure the total volume of the cell. If the test cell is directly visible, the volume can be determined accurately. If the test cell is not visible because of other cells in the tissue, make cross sections and longitudinal fresh sections at the measured depth of the microcapillary. Express the volume as the mean of the measured volumes.
6) Calculate the elastic modulus from the $d\Psi_{p(p)}/dV$ measured with injections and removals of cell solution, and the mean cell volume according to Eq. 4.7. The relation between the change in volume and the change in pressure may be S-shaped as in Fig. 4.14 because of water entry and exit through the plasmalemma at large $d\Psi_{p(p)}$. Significant water movement occurs at the extremes because of the longer times necessary to move the meniscus larger distances. Sometimes the curve is not symmetrical (Tomos *et al.*, 1981). The linear part of the curve close to the origin gives the most accurate value of $d\Psi_{p(p)}/dV$.

The largest uncertainty is the volume of the cell. Replicate cells generally have different volumes. Moreover, ε becomes larger in larger cells (Steudle *et al.*, 1977). One might expect this relation because larger cells will be required to have walls that withstand larger force (pressure has units of force per unit area). Therefore, if the volume of the probed cell cannot be directly measured, use a mean value for similarly sized cells. Calculate the variation in ε using the method of Propagation of Errors described in standard statistics texts.

CAPACITANCE

The capacitance (C) can be calculated from the measurements already described. Use the data for $\Psi_{s(p)}$, ε, and V to calculate $V/(\varepsilon - \Psi_{s(p)})$ as in Eq. 4.8. The largest uncertainty is in V and ε. Calculate the statistical uncertainty with the Propagation of Errors method described in standard statistics texts.

HYDRAULIC CONDUCTIVITY

The hydraulic conductivity (Lp) is determined by measuring the flow of water through the plasmalemma/cell wall with a pressure probe. Injecting or removing cell solution changes the water potential because the turgor changes. Water leaves or enters the cell in response (see Fig. 4.2B). The flow occurs until the cell returns to equilibrium with its environment, and the time for half of the return can be measured (half-time, $t_{1/2}$ in seconds) as in Fig. 4.2D. Using the $t_{1/2}$ and the other measurements described earlier, the Lp can be determined according to Eq. 4.9.

1) Measure the turgor in the cell as described earlier. For a tissue, note the distance the microcapillary tip has penetrated the tissue.
2) Raise and lower the turgor rapidly to determine ε as above (Figs. 4.2A and 4.2C).
3) Raise the turgor but do not lower it (Fig. 4.2B). Allow the turgor to relax as water flows out of the cell. Lower the turgor but do not raise it. Allow the turgor to relax as water flows into the cell. Determine the $t_{1/2}$ for each relaxation by noting the time for half the relaxation to occur as in Fig. 4.2D. The $t_{1/2}$ should be the same for efflux and influx.
4) Determine the Ψ_s by rapidly removing a sample of the cell solution and measuring its freezing point in a nanoliter osmometer. The Ψ_s does not vary greatly and contributes only slightly to the calculated value of the hydraulic conductivity. This step sometimes is simplified by determining the Ψ_s of a tissue extract that gives an average for the tissue.

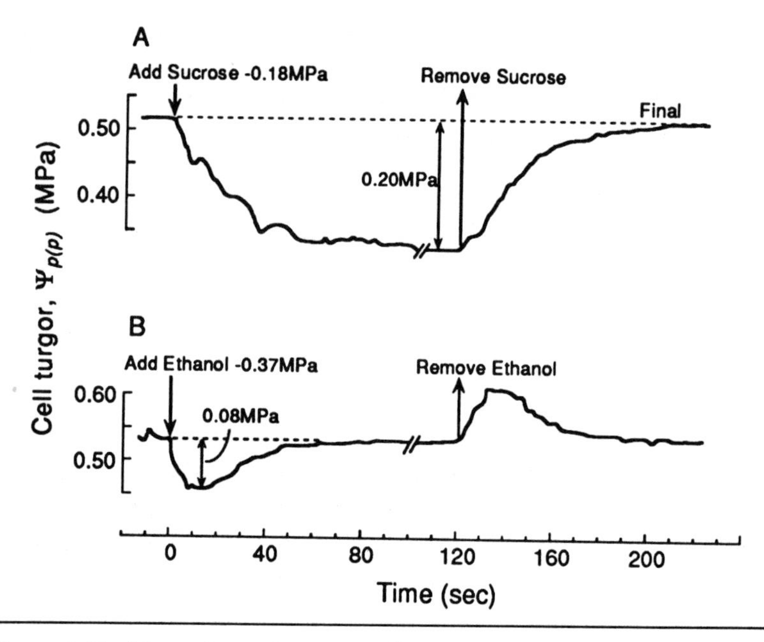

Figure 4.15. Measuring reflection coefficients σ with the miniature pressure probe. A) Sucrose (Ψ_s = -018 MPa) is placed outside a *Tradescantia* cell and the change in turgor is measured (-0.20 MPa) as the cell reequilibrates, giving a σ of 1.1. B) Ethanol (Ψ_s = -0.37 MPa) is placed outside the same *Tradescantia* cell and the change in turgor is measured (-0.08 MPa) giving a σ of 0.21. Note that only water flows out of the cell in sucrose but water flows out and solute flows in with ethanol. From Tyerman and Steudle (1982).

5) Measure the volume (*V*) and surface area (*A*) of the sampled cell from the dimensions determined microscopically. If the position of the sampled cell is not known precisely, the mean *V* and *A* should be determined at the estimated position of the microcapillary tip.
6) Calculate the *Lp* for the plasmalemma/cell wall complex of the cell according to $Lp = 0.693V/(\varepsilon - \Psi_s)At_{1/2}$ from Eq. 4.9. An example for a fairly large cell (Hüsken *et al.*, 1978) gives *V* of 11 x 10^{-12} m^3 and $d\Psi_{p(p)}/dV$ of 1.1 x 10^{11} MPa·m^{-3} so that ε is 1.2 MPa. The Ψ_s measured with extracts of cell solution is -0.25 MPa. The mean *A* determined from cell dimensions is 3.0 x 10^{-7} m^2, and $t_{1/2}$ is about 300 sec. The *Lp* calculated from Eq. 4.9 is then 5.8 x 10^{-8} m·sec^{-1}·MPa^{-1}, which is within the range of values expected for cells (Kramer and Boyer, 1995).

REFLECTION COEFFICIENT

The reflection coefficient depends on the solute and membrane and can be most simply measured by determining the change in cell water potential that is caused by solute placed on the other side of the membrane. In the equilibrium state

$$\sigma = \frac{d\Psi_w}{d\Psi_s}, \tag{4.12}$$

which indicates that solute supplied externally to change Ψ_s by 1 MPa will change the Ψ_w internally by 1 MPa when $\sigma = 1$.

Figure 4.15A shows that sucrose having Ψ_s of -0.18 MPa caused a decrease of -0.20 MPa in cell turgor in *Tradescantia* epidermal cells (Tyerman and Steudle, 1982). Assuming that the turgor change causes the same change in the water potential, this indicates that σ = essentially 1 for sucrose in this cell. Note that Fig. 4.15B for ethanol having Ψ_s of -0.37 MPa shows a water potential change of only -0.08 MPa, which gives $\sigma = 0.21$. Ethanol passes through the membrane and its osmotic effectiveness is thus only a small fraction of that for sucrose. The movement of ethanol through the membrane can be seen by the two-phase pattern of the turgor response: at first the cell shrinks as water leaves the cell more rapidly than ethanol enters but after some time water movement slows and ethanol entry predominates, causing the cell to reswell. This reswelling is diagnostic for solute entry. For sucrose, there is no sign of reswelling. Note also that the ineffectiveness of ethanol is apparent from the beginning even though the concentration difference for ethanol is largest at that stage across the plasmalemma. No matter how fast measurements are made, the osmotic effectiveness is low.

Equation 4.12 has been used to measure reflection coefficients around 1 (e.g., Tyerman and Steudle, 1982) but, for σ less than 1, the two-phase behavior of the cell causes experimental difficulties. As Tyerman and Steudle (1982) point out, permeating solutes can be dragged along by the water moving through the membrane and swept away from the membrane surface. There are unstirred layers of water and solute next to the membrane and these can limit solute and water transfer. The results depend on how fast the solute penetrates the membrane. Thus, a σ below 1 clearly indicates that the membrane is nonideal but the actual value of σ is usually approximate.

Precautions

VIBRATION

Even moderate vibration can cause pressure probe measurements to fail. Test for vibration by observing the tip of the microcapillary at 80X with the stereomicroscope. If the tip is not clear but rather appears blurred, and if it becomes clearer when it touches a heavy immobile surface, there is too much vibration. The clarity of the free probe should be the same as that of the probe immobilized against the surface. Test whether the source of vibration is air movement by constructing a container to cover the probe, thus preventing the air from stirring.

If the problem is not caused by air movement, it likely is caused by the vibration of the floor on which the probe bench sits. Locate the probe close to a part of the room having a structural support for the building or, better yet, work on a floor that is directly in contact with the ground.

Further stability can be achieved with a vibration-damping table. Electronic and pneumatic designs work well and operate in particular frequency ranges. If the frequency of vibration is unknown, obtain a steel plate about 2 cm thick and place it on some foam rubber or similar vibration-damping material. Mount the probe on the steel plate. Be sure to make the plate large enough to hold the sample and the microscope so that all movement will be as a unit.

EVAPORATION

Evaporation is induced because of the need to illuminate the probe tip and meniscus inside the microcapillary. The illumination causes local heating of the tissue and evaporation even when the surrounding air is saturated with water vapor. The loss in turgor is most apparent in surface cells but can affect cells deep inside the tissue.

The control of evaporation is one of the most important principles of pressure probe use. The degree of saturation should not be tested with a humidity detector because most do not give reliable readings around 100% humidity. Instead hang wet filter paper in the air and determine the weight loss. If there is any weight loss, the air is not saturated. Even small sources of dry air can decrease humidities sufficiently to allow weight loss. Sometimes these are minimized by saturating the air in the room, then saturating the air in a small chamber around the probed plant part.

Saturating the air will eliminate evaporation only if the evaporating surface has the same temperature as the air. Above air temperature, evaporation can be rapid. To avoid this problem, the tissue can be coated with petrolatum (Vaseline) to create a vapor barrier. Another approach is to use minimal light to illuminate the sample, thus reducing the temperature of the tissue.

TEMPERATURE

The probe and oil expand and contract with changes in temperature. To avoid temperature effects, the instrument should not be exposed to large thermal gradients or rapid fluctuations in temperature. The temperature control of most laboratories is stable enough to allow the probe to be used but you should avoid air ducts, direct sunlight, nearby compressors, and so on.

Temperature also affects the pressures in cells because the osmotic potential varies with temperature, as discussed in Chap. 1 and shown in Appendix 3.2 for sucrose solutions, and turgor responds accordingly. However, the effects are not large because the osmotic potential responds to the Kelvin temperature (see Kramer and Boyer, 1995).

PLUGGED TIP

Most measurements with the pressure probe are made in the central vacuole of the cell. Cells that have a small vacuole or a thick layer of cytoplasm can cause plugging of the tip of the microcapillary with cytoplasm. A plugged microcapillary is detected from the extreme changes in pressure that occur when the metal rod is moved in or out of the silicone oil, and from the lack of response of the meniscus to the pressure change. It is essential to test for an open tip during a measurement because the pressure in a plugged tip can be mistaken for the turgor pressure of the cell but bears no relationship to the turgor (see Turgor Pressure, step 4, and Fig. 4.12).

Sometimes the tip can be unplugged by raising the pressure and blowing the plug away. With large cells, it is sometimes possible to work with a tip having a larger opening. When this is not possible, the tip should be moved to another cell. If plugging persists, it may be necessary to replace the microcapillary.

REPLACING SEALS

Instrument leaks are usually caused by faulty rubber seals. If a test of instrument leaks fails (see Leaks), install new seals and repeat the test. New seals can be made from a sheet of rubber gasket material of low to intermediate hardness. Obtain a punch of the diameter required and cut several disks. Place the disks in a freezer at -80°C or on dry ice or in liquid N_2 to harden the rubber, then rapidly drill a hole in the center to admit the microcapillary, metal rod, or other part of the pressure probe. The rubber should remain hard as long as it is cold. The hole should be the same diameter as the part passing through the seal (1 mm for microcapillary) or slightly smaller.

After several years of use, cracks may develop in the main body of the probe because of pressures in the borings. The cracks will gradually work their way to the outside as the probe is used, and a leak results. These cracks will be visible and their progress can be followed. As the crack nears the surface, the probe body will need to be replaced with a new one.

STATISTICS

Measurements of turgor among individual cells of a tissue typically vary by ±0.1 MPa or more (see Fig. 5.2), and single cells may show a similar variation. It is essential to measure a minimum of 7-10 cells for a statistically relevant sample of the turgor. A similar sample size is needed for measuring cell dimensions. The mean ±1 standard deviation is the usual statistical test, and for reporting the mean of several means of 7-10 cells each, the standard error is used.

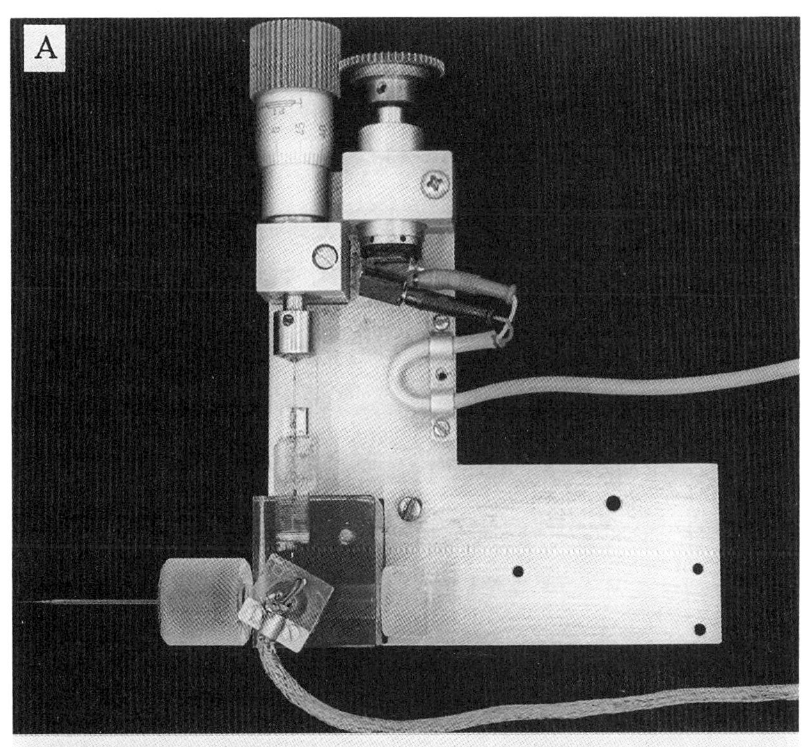

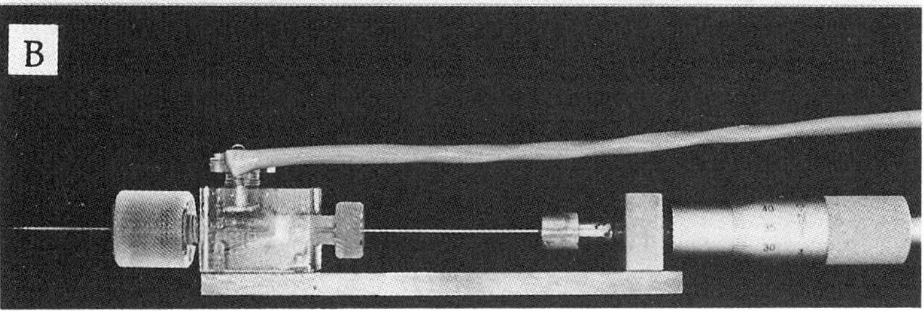

Figure 4.16. A) Miniature pressure probe and B) pressure probe for large cells.

Appendix 4.1-Building a Pressure Probe

The pressure probe consists of a block of Plexiglas that has been bored to allow silicone oil to move between the microcapillary, metal rod, and pressure transducer. The pressure probe can be made for cells of various sizes by changing the diameter of the borings in the block. For the smallest probe, the borings have a diameter of 0.5 mm which is suitable for the smallest cells. For larger cells, borings are large to accommodate the larger volumes of liquid that are moved in and out of the cell. Figure 4.16 shows a detailed view of the miniature probe and a larger one suitable for large internodal cells of *Chara* or *Nitella*. The miniature probe uses a motor to drive the metal rod and thus minimizes vibration from handling the micrometer head. The large probe allows the metal rod to be driven by hand.

Upon settling on a design for the borings, such as shown in Fig. 4.17, have the block of Plexiglas machined and polished so that the borings can be easily seen from the outside. Mount the block on black paper on a metal plate to allow air bubbles to be easily seen in the silicone oil when filling. Mount the metal rod, micrometer head, and motor on the metal plate as shown (Fig. 4.16 and 4.17). Mount this assembly on a micromanipulator (Fig. 4.4). For various parts for the probe, see Table 4.1.

The power supply provides 10 V DC to the pressure transducer mounted on the Plexiglas block (Fig. 4.18A). The transducer returns a small DC voltage (60-70 mV/MPa) that can be detected with a strip

Figure 4.17. Construction details for a miniature pressure probe shown at a scale of 1:1. Top view extends to the plane of the microcapillary but does not show the transducer. Side view extends to the microcapillary and shows the transducer. The body of the pressure probe is made of a Plexiglas block (A) that is mounted on black paper on a metal plate. The pressure transducer (B) is sealed with epoxy (stippled) in a Plexiglas mounting (C) on the upper side of the block. An O-ring (D) in a recess (E) seals the mount in the block. The metal rod (F) and microcapillary mount on the sides of the block. The seals are punched out of rubber sheet (H) and are drilled to the size of the metal rod or microcapillary after hardening the rubber at low temperature. Two screws (I) hold the block on the metal plate. A set screw (J) holds the metal rod (F) to the micrometer screw. Metric dimensions are shown in millimeters or in millimeter diameter x number of threads per millimeter (M__x__). Not shown are the micrometer screw, motor drive, or power supply and readout.

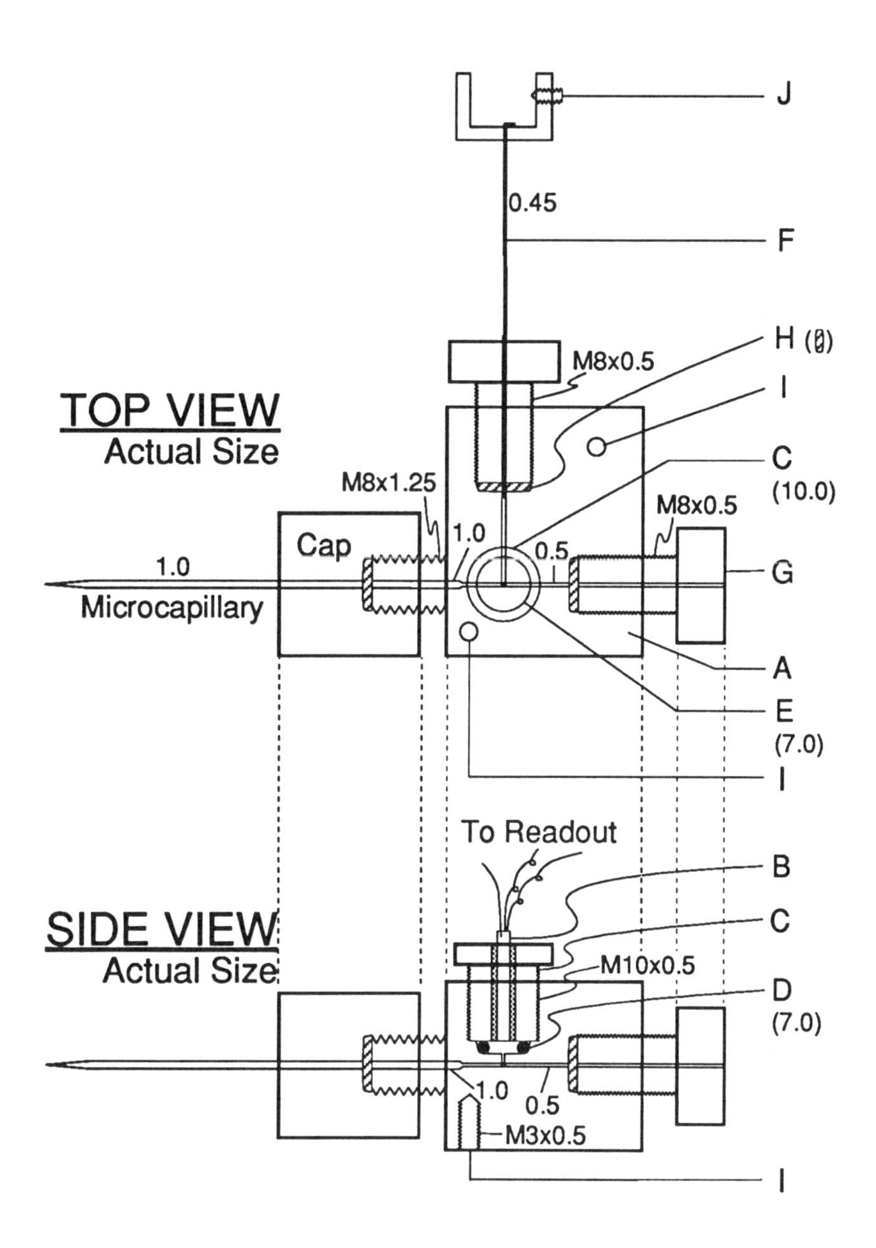

J
0.45
F
H (∅)
M8x0.5
I
TOP VIEW
Actual Size
C
(10.0)
M8x1.25
M8x0.5
Cap
1.0
0.5
1.0
G
Microcapillary
A
E
(7.0)
I
To Readout
B
SIDE VIEW
Actual Size
C
M10x0.5
D
(7.0)
1.0
0.5
M3x0.5
I

A Transducer Power Supply

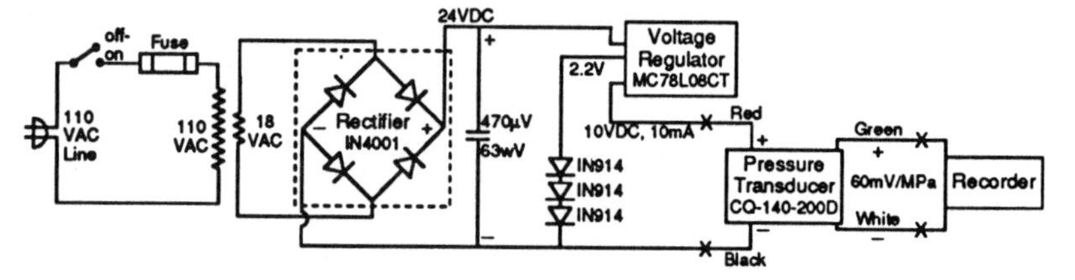

B Motor Power Supply

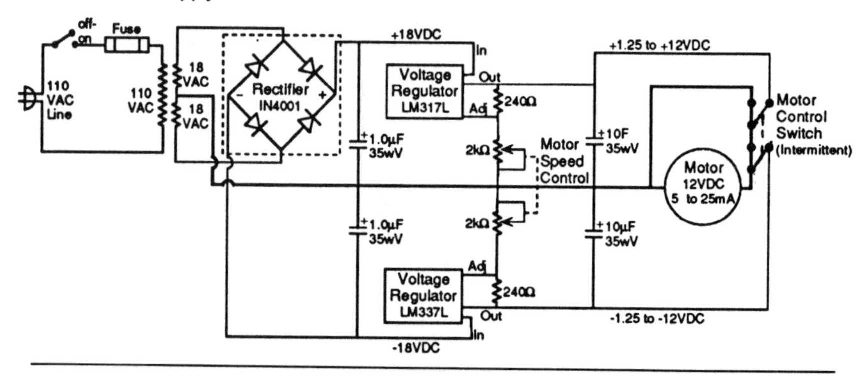

Figure 4.18. A) Power supply for the pressure transducer on the pressure probe. The transducer requires 10 VDC and consumes about 10 mA from the power supply to give an output of about 60-70 mV/MPa (see Fig. 4.6). B) Power supply for the drive motor for moving the metal rod (required for miniature probe, see Fig. 4.16). The speed control is used to vary the voltage supplied to the motor from ±1.25 VDC to ±12 VDC and the intermittent motor control switch reverses the voltage to turn the motor forward or backward. The motor consumes as much as 25 mA.

chart recorder (Fig. 4.6). Power consumption is small (about 10 mA) so that an inexpensive power supply can be used. In some systems, the DC output of the transducer is amplified so that a digital meter readout can display the pressure (not shown).

The drive motor for the metal rod in the small transducer uses 0-12 V DC and can be driven forward or backward by reversing the polarity of the voltage. The power consumption of the motor is about 25 mA at 12 V (Fig. 4.18B). The speed of the motor is varied by altering the voltage supplied. The power supply of Fig. 4.18B is variable and the voltage is adjusted with a dual potentiometer (motor speed control).

Table 4.1. Parts Manufacturers

Part	Manufacturer	Part No.	Recent Cost
Differential Pressure Transducer	Kulite Semiconductor Products, Inc. 1039 Hoyt Ave. Ridgefield, NJ 07657 (201) 461-0900	CQ-140-200-D	$466.00
Motor (12 VDC), Gear, Micrometer (Motor Mike)	Oriel Corporation 250 Long Beach Bvd. Stratford, CT 06497 (203) 377-8282	18040	$398.00
Micromanipulator	E. Leitz, Inc. Rockleigh, NJ 07647 (201) 767-1100	Leitz micromanipulator	$2500.00
Vertical Pipette Puller	David Kopf Instruments 7324 Elmo St. P.O. Box 636 Tujunga, CA 91042 (818) 352-3274	Model 720	$2695.00
Microcapillaries (thin-walled, 1.0 mm OD)	World Precision Instruments 175 Sarasota Ctr. Blvd. Sarasota, FL 34240 (813) 371-1003	TW100F-4	$50.00
Silicone Oil	Stauffer-Wacker Corp. 3301 Sutton Rd. Adrian, MI 49221 (517) 264-8500	AS-4	Free sample
Small Tubing for Filling Probe with Oil, 6" length, 0.016" OD	Small Parts, Inc. 13980 NW 58th Ct. P.O. Box 4650 Miami Lakes, FL 33014 (305) 557-8222	Q-HTX-27TW	$1.00

Table 4.1. Parts Manufacturers (continued)

Part	Manufacturer	Part No.	Recent Cost
Microcapillary Beveler	Optical Apparatus Co. 136 Coulter Ave. Ardmore, PA 19003 (800) 648-6622	EG-4 Narashige tip grinder	$1181.00
Microscopes	Optical Apparatus Co. 136 Coulter Ave. Ardmore, PA 19003 (800) 648-6622	Optiphot: Nikon Compound Microscope with reticle, long focal length 40x objective and 10x eyepiece	$2500.00
		SMZ-2T: Nikon Stereomicroscope with reticle, 2X objective and 15X eyepieces to give magnification of 30-189X	$2500.00
Nanoliter Osmometer	Clifton Technical Physics Box 181 Hartford, NY 12838 (518) 632-5260	Clifton Nanoliter Osmometer	$3000.00
Immersion Oil for Osmometer Wells	Fisher Scientific 50 Valley Stream Pkwy. Malvern, PA 19355 (800) 766-7000	12-370B: Cargille Immersion Oil, Type B, 4 oz.	$10.00
		12-370-1B: Cargille Immersion Oil, Type NVH, 4 oz.	$10.50

Chapter 5

Measuring the Water Status of Plants and Soils: Some Examples

In order to help the reader design experiments for measuring the water status of plants and soils, this chapter describes a few examples that illustrate in a practical way the kinds of information that can be gained. The examples are taken from the original literature and represent only a partial sampling. For a fuller account the reader is directed to Kramer and Boyer (1995).

Practical Benefits of Thermodynamic Equilibrium

In the previous chapters, emphasis was placed on measuring the water status at thermodynamic equilibrium. The practical benefit can be seen in comparative studies of sorghum leaves made by De Roo (1969) using equilibrium methods and Blum *et al.* (1973) using similar methods but not operated at equilibrium. De Roo (1969) compared the water potential measured with an isopiestic psychrometer and the xylem pressure measured at balance in a pressure chamber and found them to be virtually equivalent, i.e., close to the line of equivalency shown in Fig. 5.1A. De Roo (1969) also determined the osmotic potential of the xylem solution and found it to be on average -0.05 MPa. Adding this potential to the xylem pressure in Fig. 5.1A gives the leaf water potential (see Chap. 2) and an even closer correspondence to the equivalency line. Note that there was little variation in the data. Because the data fall on the line of equivalency, there is no uncertainty about which method is correct and no need to calibrate. The pressure chamber compares so well with the psychrometer, which has already been shown to give absolute values of the water potential (see Chap. 3), that both methods can be considered to give absolute values of the water potential of sorghum leaves.

Blum *et al.* (1973) also studied sorghum leaves but in contrast to De Roo (1969) used the pressure chamber as a nonequilibrium method by noting the first appearance of xylem solution when the sample was pressurized at a constant rate (Figs. 5.1B and 5.1C). A comparison was made with thermocouple psychrometer readings by the

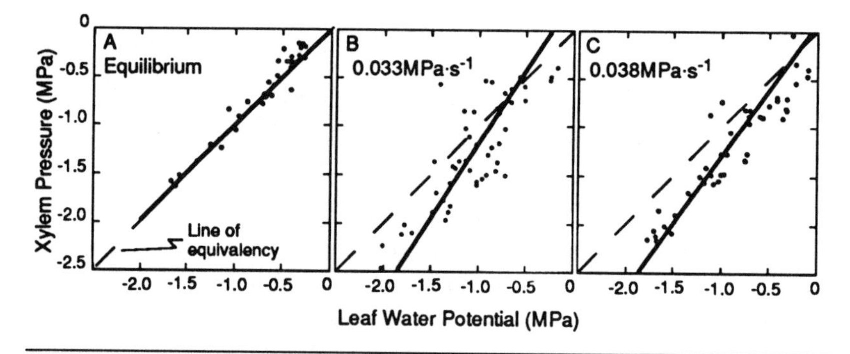

Figure 5.1. Comparison of the water status of sorghum leaves measured with a psychrometer and a pressure chamber. A) Leaf water potential determined with an isopiestic thermocouple psychrometer and compared with the xylem pressure at balance in the same leaf. These are equilbrium methods. The line of equivalency (dashed line) indicates a 1:1 correspondence between them. Data from De Roo (1969). B) The same comparison using a Peltier psychrometer for leaf water potential and a pressure chamber pressurized at a steady rate (0.033 MPa·sec^{-1}) for xylem pressure. These are not equilibrium methods. Data from Blum *et al.* (1973). C) As in B except the pressure chamber was pressurized more rapidly (0.038 MPa·sec^{-1}). Data from Blum *et al.* (1973).

Peltier method which also is not an equilibrium technique. It is clear that the data do not match the line of equivalency and that the relation depends on the rate of pressure application (cf. Figs. 5.1B and 5.1C). Also, the data show a large variability. Blum *et al.* (1973) indicate that if one uses nonequilibrium methods, careful calibration is essential. They also point out that it is not clear whether the pressure chamber or psychrometer gives the more accurate values. Therefore, one may consider the nonequilibrium data to be only relative approximations. Clearly, the equilibrium techniques used by De Roo (1969) are preferred. Their freedom from calibration and lack of variability are desirable features, and the ability to interpret the measurements is simplified by having absolute values of the potential.

Large Tensions Demonstrated in the Apoplast

Scholander *et al.* (1965) interpreted their pressure chamber measurements as a demonstration that large tensions exist in the xylem and apoplast of plants. Large tensions are easily disrupted because water is in a quasi-stable state and water columns can break because of cavitation. As a consequence, doubts are sometimes expressed that such large tensions can exist, and indeed attempts to demonstrate large tensions sometimes fail (see Kramer and Boyer, 1995 for some recent examples). This is especially true when the xylem is penetrated with a pressure probe because the penetration itself can disrupt the water column by causing cavitation. On the other hand, the psychrometer measures the water potential directly in the apoplast and does not penetrate the water column. As shown by De Roo (1969), the osmotic potential of the apoplast solution was only -0.05 MPa in sorghum and the only other possible potential is a tension (part of $\Psi_{m(a)}$ in Eq. 2.4 and Fig. 2.4B). Therefore, the low water potentials in the apoplast (Fig. 5.1A) are strong evidence supporting the interpretation that large tensions exist.

It should be noted that the isopiestic psychrometer is the only method shown to give accurate absolute values of the water potential using plant tissue of known water potential (see Fig. 3.7). Leaves on intact plants have water potentials similar to the water potentials of excised samples from the same leaves (Boyer, 1968). Therefore, excision does not alter the conclusions, and the tensions have been demonstrated in many species using the isopiestic psychrometer and the pressure chamber (Boyer, 1967a; Ghorashy *et al.*, 1971; De Roo, 1969).

Single Cell and Tissue Measurements Compared

Figure 5.2 compares the variation in turgor in cells measured with a miniature pressure probe and in tissues grown identically and measured with an isopiestic psychrometer (Nonami *et al.*, 1987). Mature tissue was used from stems of soybean seedlings after transplanting to vermiculite of low water content, and it is apparent that the turgor initially decreased (Fig. 5.2A), then increased as recovery occurred without rewatering when the roots reconnected with the water supply (Fig 5.2B).

The variability was greater among individual cells (pressure probe) than among the tissue samples (psychrometer). For the cells, the variation included differences among plants and among cells of the same plant, and perhaps some variation attributable to the probe. For the tissues, many cells contributed to each tissue sample and four to six

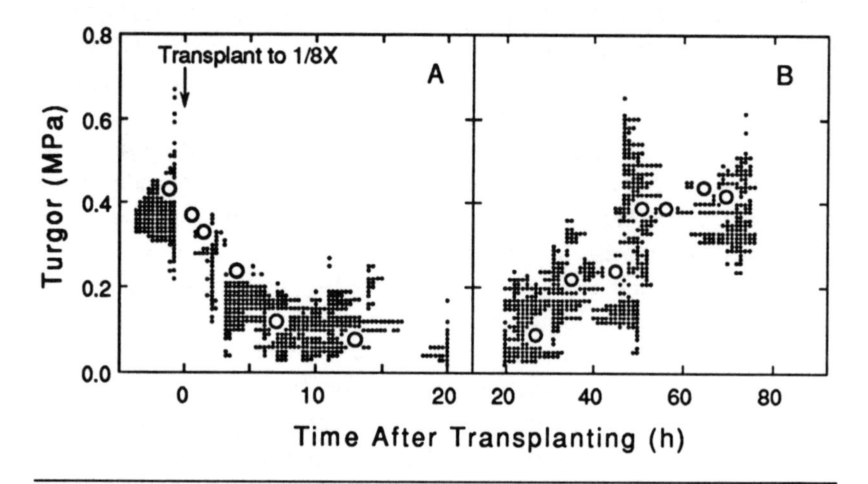

Figure 5.2. Turgor pressures measured in the mature region of soybean stems using a miniature pressure probe (small points) and an isopiestic thermocouple psychrometer (open circles). Pressure probe data are for individual cells of the intact plant, psychrometer data are for excised tissue segments. A) The soybean seedlings were transplanted to vermiculite of low water content at 0 hr whereupon turgor fell for 15 hr. B) Twenty hours after transplanting, turgor recovered because the roots reconnected with the water supply. Each data point is an individual measurement. From Nonami *et al.* (1987).

plants were represented in each measurement to give an average. This illustrates that there is no substitute for individual cell data when one is exploring cell properties, but there is less labor with the psychrometer (or pressure chamber) when measuring the water status of tissues. The psychrometer gives simultaneous averages for all the cells, which is not possible with the pressure probe.

It is noteworthy that the two methods agree because they rely on different principles of measurement. The agreement gives confidence in the results, and it appears that penetrating the cells with the microcapillary and excising the mature tissue for the psychrometer do not disturb the turgor significantly. A comparison of tissue turgor also has been made with sunflower leaves using a pressure chamber at balance and an isopiestic psychrometer (Boyer and Potter, 1973) and gave similar results provided a correction was made for dilution of the cell solution by apoplast solution during the measurement with the

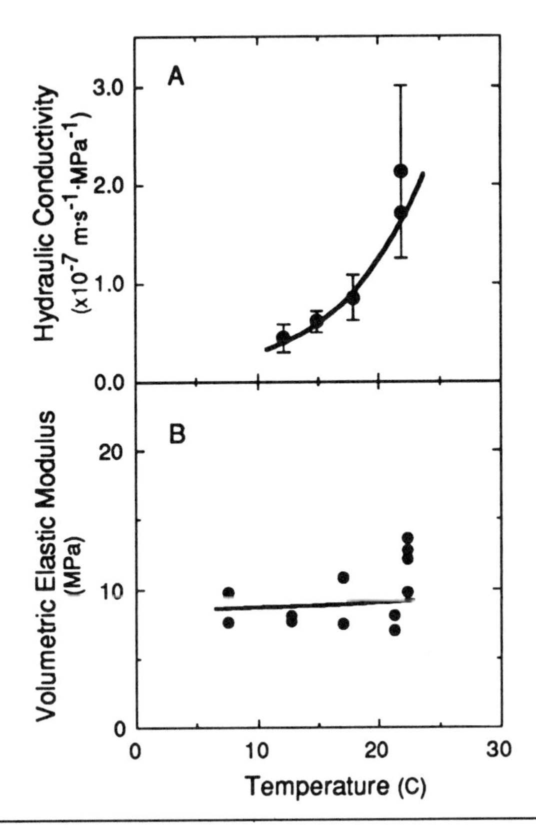

Figure 5.3. Temperature response of (A) hydraulic conductivity and (B) volumetric elastic modulus in *Tradescantia* epidermal cells. Measurements were made with a miniature pressure probe. Data from Tomos *et al.* (1981).

psychrometer (see Chap. 3). With the correction, the osmotic potentials were similar for the two methods (Boyer and Potter, 1973).

Temperature, Membrane Transport, and Cell Walls

Biological activity depends on enzyme reactions that are markedly responsive to temperature, and the movement of ions and metabolites across membranes varies with temperature in part because of the involvement of enzyme reactions in their transport. By contrast, water moves passively into cells, driven by differences in potential on the two sides of the plasmalemma, as can be demonstrated readily with a pressure chamber or pressure probe. For strictly passive transport through a membrane, water would be expected to move slightly more rapidly as temperatures rise and the viscosity of water decreases moderately. However, Fig. 5.3A shows that the hydraulic conductivity of *Tradescantia* cells increases about 5-fold as the temperature rises to 22°C from 12°C (Tomos *et al.*, 1981) and is more like the thermal response of an enzymatic reaction than the viscosity, which changes only 1.3-fold. A possible explanation is that pore-forming proteins appear to be involved in water transport (Kramer and Boyer, 1995), and protein synthesis and insertion into the membrane might influence membrane conductivity because these processes are highly responsive to temperature.

Likewise, the cell wall is synthesized by metabolic reactions that are temperature-dependent but the wall elasticity is a physical property whose temperature response is difficult to predict. Tomos *et al.* (1981) showed that the volumetric elastic modulus of the wall is virtually constant at various temperatures in *Tradescantia* (Fig. 5.3B). The modulus is used to calculate the hydraulic conductivity of membranes for water (see Chap. 4), but this result indicates that the modulus is not the cause of the large conductivity response to temperature.

Importance of Growth

Plants grow mostly by enlarging the cells produced in meristems. The enlargement requires water uptake and some solute transport, and excising the tissues disrupts both processes. The psychrometer for intact plants does not suffer from this problem and is the only method that completely maintains the integrity of tissues and cells (see Chap. 3). The method has been used to study the growth process (Boyer, 1968; Boyer *et al.*, 1985). In intact plants, the water potential of the growing tissue is typically below that of the nearby mature tissue when no transpiration is occurring (Boyer *et al.*, 1985; Cavalieri and Boyer, 1982; Matyssek *et al.*, 1991a). As shown in Fig. 5.4, when the growing tissue is excised, the water potential rapidly decreases by a small amount (Boyer *et al.*, 1985; Matyssek *et al.*, 1991b) because of continued expansion of the walls which decreases (relaxes) turgor after the water and solute flows are disrupted (Boyer *et al.*, 1985; Cosgrove, 1985, 1987; Cosgrove *et al.*, 1984; Matyssek *et al.*, 1988).

The relaxation can be observed over a range of temperatures and is similar as long as some growth is occurring (Fig. 5.4). However,

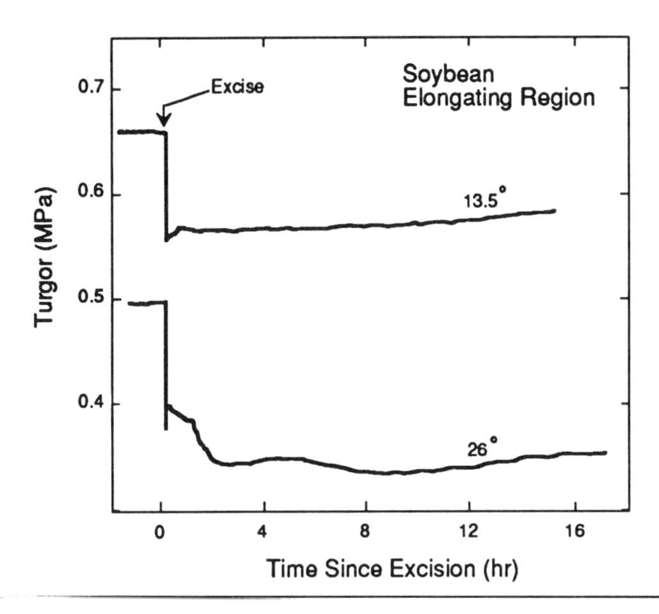

Figure 5.4. Turgor decreases when the stem growing region is excised at temperatures favoring rapid growth (26°C) or slow growth (13.5°C) in soybean seedlings. Turgor initially was measured in an isopiestic thermocouple psychrometer for intact seedlings, and excision occurred at the arrow without disturbing the sample being measured. The stems were growing at the time of excision. Because the excision prevented all water from entering the enlarging cells, the turgor dropped rapidly (5 min) as the walls relaxed after the excision until a threshold turgor was reached below which no further relaxation occurred. Note that the rapid relaxation was similar at the differing temperatures but the growth rate at 13.5°C was only about 5 to 10% of that at 26°C. From Boyer (1993).

the relaxation causes turgor to decrease only about 0.1 MPa (Fig. 5.4) and thus relaxation often cannot be seen within the natural variation between samples. Nonami and Boyer (1989) compared four methods of measuring turgor in soybean stems known to show the relaxation behavior of Fig. 5.4. The data for intact plants in the psychrometer (Fig. 5.5A) did not differ significantly from the data in the excised tissue psychrometer (Fig. 5.5B), the pressure chamber (which also requires excised tissue, Fig. 5.5C), or the miniature pressure probe, which had to penetrate individual cells (Fig. 5.5D). Thus, in many studies, wall relaxation does not significantly affect the measurements.

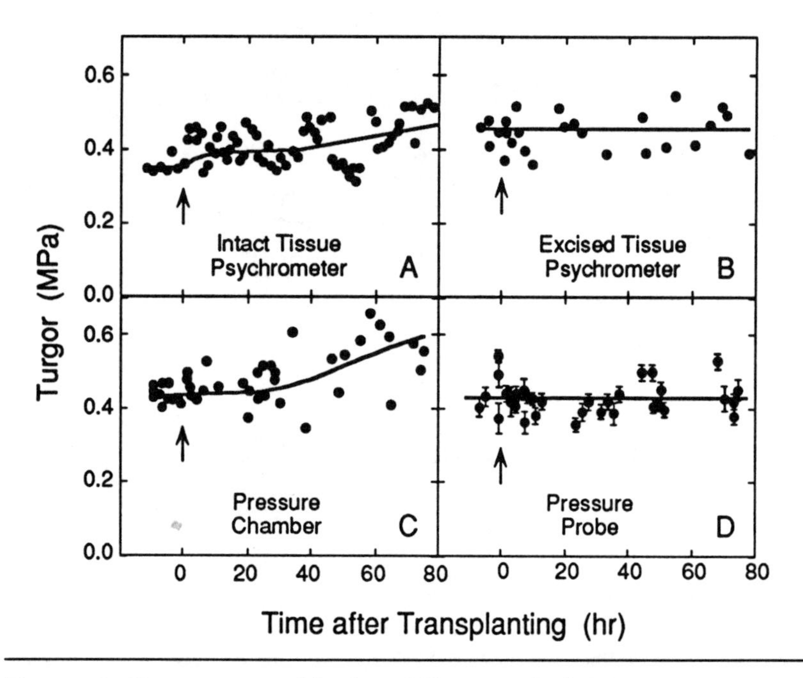

Figure 5.5. Turgor measured by four different methods in growing tissues of soybean stems before and after transplanting to vermiculite having a low water content. The vermiculite water potential at high water content was -0.01 MPa, and at low water content was -0.3 MPa measured with an isopiestic thermocouple psychrometer. Transplanting occurred at the arrow and inhibited growth to near zero within 2 hr. A) Isopiestic psychrometer for intact plants. B) Isopiestic psychrometer for excised tissue. C) Pressure chamber which required excised tissue. D) Miniature pressure probe which penetrated individual cells of intact plants. The turgor for methods A, B and C was measured from the difference between the water potential and osmotic potential. The turgor for method D was measured directly. Data from Nonami and Boyer (1989).

The relaxation sometimes is not seen (e.g., Westgate and Boyer, 1984), usually because it can be delayed for hours if mature or slowly growing tissue is excised with the growing tissue (Boyer *et al.*, 1985; Matyssek *et al.*, 1988). This is because the attached mature tissue acts as a water source that prevents relaxation until the source is depleted (Matyssek *et al.*, 1988, 1991a,b). Previous claims (Cosgrove, 1985; 1987;

Cosgrove *et al.*, 1984) that wall relaxations were large were artifacts caused by these delays followed by large turgor changes (Matyssek *et al.*, 1988).

These experiments indicate that growth can affect measurements of plant water status but, as long as care is taken to ensure that the measurements are stable and made rapidly, they reflect either a small relaxation or no relaxation. Either case is acceptable in practical terms because the effects are generally smaller than the inherent biological variability between samples. For detailed studies of the mechanism of growth, however, knowing the extent of wall relaxation has proven valuable (Kramer and Boyer, 1995).

Growth-Induced Water Potentials

The lower water potential in growing tissue than in mature tissue is growth-induced because it disappears when growth is inhibited by low temperature (Boyer, 1993) or by auxin depletion (Maruyama and Boyer, 1994). Figure 5.6 shows that, in rapidly growing stem internodes of maize, growth was rapid at the base of the internode but not at the top, and the water potential was significantly lower in the base than at the top. This growth-induced water potential did not appear in internodes that were uniformly mature (Fig. 5.6). The measurements were made at the end of the night period when transpiration was negligible and water transport was for growth alone. Similar growth-induced water potentials could be seen in all the growing tissues of the plant (Westgate and Boyer, 1984, 1985) as long as transpiration was minimal.

Plants such as maize are favorable for these kinds of studies because, with the exception of some of the reproductive tissues, all the growing regions are enclosed by other tissues or by soil, and transpiration does not occur directly from the growing tissues. Suppressing transpiration for the whole plant allows the xylem to be at a uniform water potential so that surrounding mature tissues can be compared with growing tissues.

Just as observed with the water potential, the turgor was lower in the growing tissue than in the mature tissue (Fig. 5.6). This indicates that some factor prevented turgor from becoming as high as in nongrowing cells. It appears that the enlargement of the cell wall is

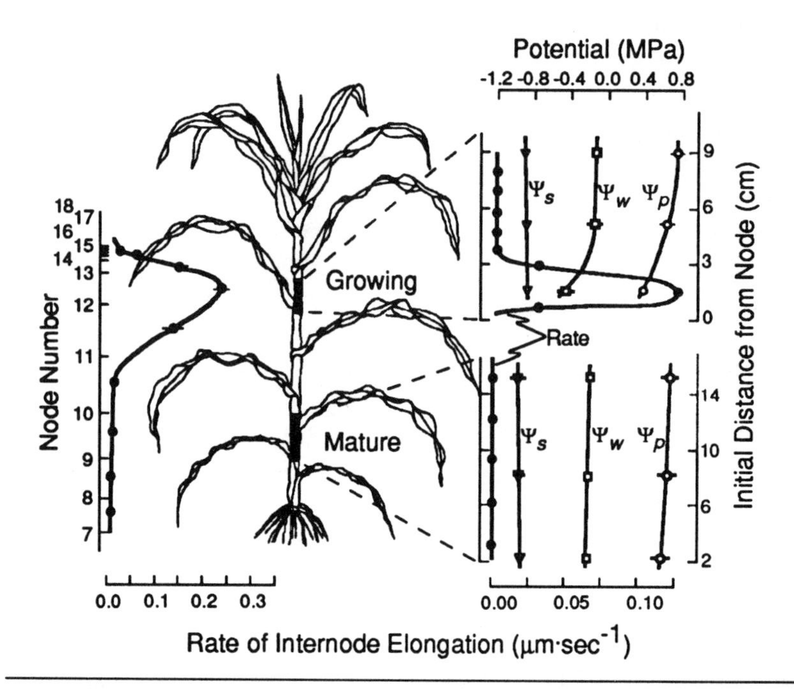

Figure 5.6. The growth and water status of maize stems in predawn conditions when transpiration was negligible. (Left) The distribution of growth along the stem. (Right upper) The distribution of growth within a rapidly growing internode in the upper stem and the water potential (Ψ_w), osmotic potential (Ψ_s), and turgor (Ψ_p) at various positions. (Right lower) As in the right upper graph but for a nongrowing internode at the stem base. The water status was measured with an isopiestic psychrometer using excised tissues. The soil Ψ_w also was measured and was about -0.04 MPa. The mature tissues had a Ψ_w close to that in the soil indicating that the xylem also had a similar water potential. Therefore, the low Ψ_w and Ψ_p of the growing tissues in the upper right graph occurred outside of the xylem in the growing cells. Data from Westgate and Boyer (1984).

rapid enough to prevent a complete buildup of turgor (Boyer, 1968, 1993; Nonami and Boyer, 1987; Maruyama and Boyer, 1994), and the water potential is lowered and transmitted to the apoplast as a tension measurable with the pressure chamber (Nonami and Boyer, 1987). The tension pulls water into the enlarging cells from the xylem. The tension

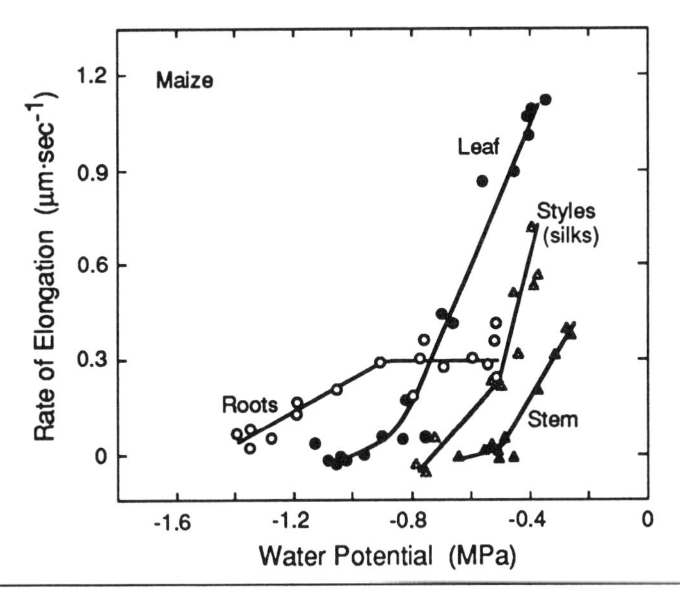

Figure 5.7. Growth of leaf, root, stem, and styles (silks) of maize at various water potentials in the growing region of each organ. Plants were grown in soil from which water was withheld. Water potentials were measured in excised tissue with an isopiestic thermocouple psychrometer at the end of the 10 hr night period when transpiration was negligible. Growth was measured during the entire preceding 10 hr. Data from Westgate and Boyer (1985).

increases until water moves into the cells at a rate that satisfies the demand of enlargement.

Growth at Low Water Potentials

As water potentials decrease, the growth rate often slows. Figure 5.7 shows that maize leaves grew less when water was withheld from the soil sufficiently to decrease the water potential of the growing region at the leaf base (Westgate and Boyer, 1985). The roots of the same plants showed little response until water potentials became very low at the growing root tips. This differential response kept the roots growing while retarding the growth of the leaves and had obvious advantages for a plant coping with dehydrating soil. The growth of the styles (silks) and stem was even more inhibited than in leaves. As the

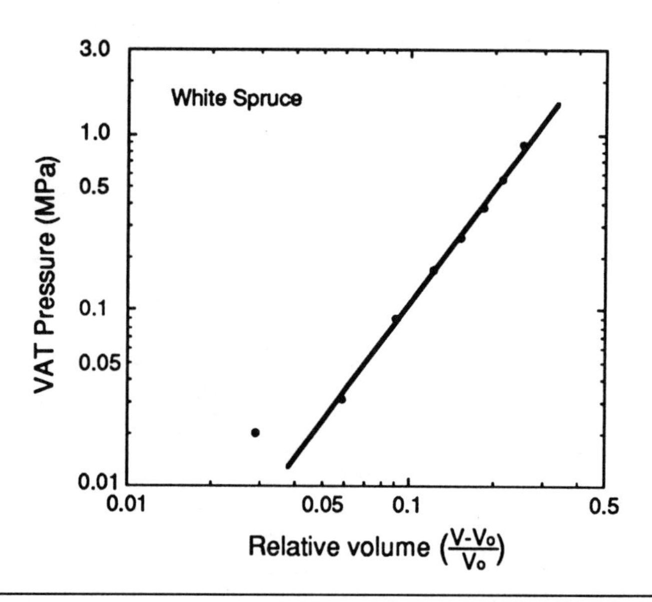

Figure 5.8. Volume-averaged turgor (VAT) at various tissue water contents measured with a pressure chamber in white spruce twigs. The tissue water content is expressed as the amount of water volume V that exceeds the volume at incipient plasmolysis V_0 (zero turgor) as a fraction of V_0. Data from Tyree and Hammel (1972).

enlargement of the silks and stem is necessary for flowering in maize, reproductive development was severely retarded by water deprivation. The cause of this differential response is not fully understood but is clearly regulated internally because the water potential was measured in the growing part of each organ, and differences at a given water potential can only be explained by factors within the plant.

Turgor Measured with a Pressure Chamber

When Tyree and Hammel (1972) first showed that the pressure chamber measured the volume-averaged turgor (VAT) in plant tissue, they could easily measure the turgor over the whole range of tissue water contents with a single sample. They observed that a plot of the logarithm of the turgor versus the logarithm of the sample water content (volume) gave a linear relation (Fig. 5.8 shown for white spruce). Although this is an empirical relationship whose mechanism remains unknown, the linearity held for all the species in the experiment, and the slopes of the lines while different were not widely different. As expected, the line for each species was displaced relative to the others reflecting the difference in water content at which the tissue attained a particular turgor because of the differences in the cell wall elasticity among various species. However, the linearity of the relationship suggests that almost any turgor can be predicted for any water content if the turgor is known at only two points.

Varietal Differences in Midday Water Potential under Field Conditions

Each day, plants undergo changes in water status that reflect the forces required to extract water from the soil. Depending on the frictional resistance of the water path between the leaves and the soil, the water potential of the leaves will need to be higher or lower. Agricultural crops yield best when the water potentials do not become too low during midday because maximum photosynthesis occurs at that time and low water potentials can mean losses in photosynthesis and growth. To prevent leaf water potentials from decreasing to inhibitory levels, root and vascular development must be sufficient to keep the frictional resistance at a moderate level and supply adequate water to the leaves. However, too much root and vascular tissue can be deleterious because it uses dry matter that otherwise could be devoted to marketable yield. Therefore, geneticists and breeders fit plants genetically to the water demand of the region where the plants will be grown. This is usually done with yield trials that integrate all the yield factors in a particular region, automatically selecting for optimum frictional resistance for water flow.

We can see how this occurs by comparing the midday leaf water potentials of older cultivars with those of newer cultivars of soybean when the plants are grown together in the same field (Boyer *et al.*, 1980). Because the cultivars have a common water source, differences in leaf water potential indicate differences in frictional resistances among the cultivars. Figure 5.9 shows that Wayne, a newer cultivar, had a leaf water potential close to -1.2 MPa in midday but the other cultivars had lower water potentials measured in the field with

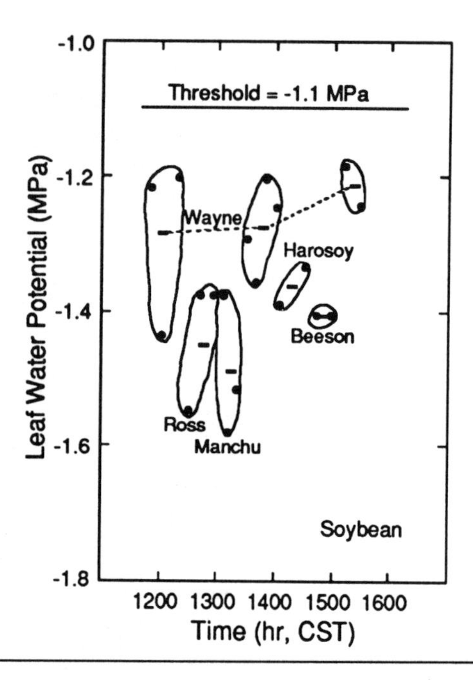

Figure 5.9. Leaf water potential measured with a pressure chamber during the afternoon in various cultivars of soybean growing in the same field in Urbana, Illinois. Closed circles are individual measurements and the bars are the means for each cluster of measurements. The measurements were made only in (1) fully exposed leaves perpendicular to the incoming radiation, (2) at the top of the canopy, (3) when reference water potentials (in Wayne) were stable, and (4) in a portable pressure chamber standing next to the plant to ensure minimal water loss after excising the leaf. The threshold water potential of -1.1 MPa indicates the potential below which photosynthesis is inhibited. Data from Boyer *et al.* (1980).

a pressure chamber. In soybean, photosynthesis is inhibited when leaf water potentials are below -1.1 MPa (Boyer, 1970; Ghorashy *et al.*, 1971; Huang *et al.*, 1975), and the water potentials exhibited by the various cultivars could inhibit photosynthesis as much as 50%.

Figure 5.10 shows that, among soybean cultivars in Maturity Group II where Beeson is the newest cultivar (introduced in 1968) and the older cultivars are part of the Beeson lineage, the yield increased as newer cultivars were released by plant breeders but the midday water

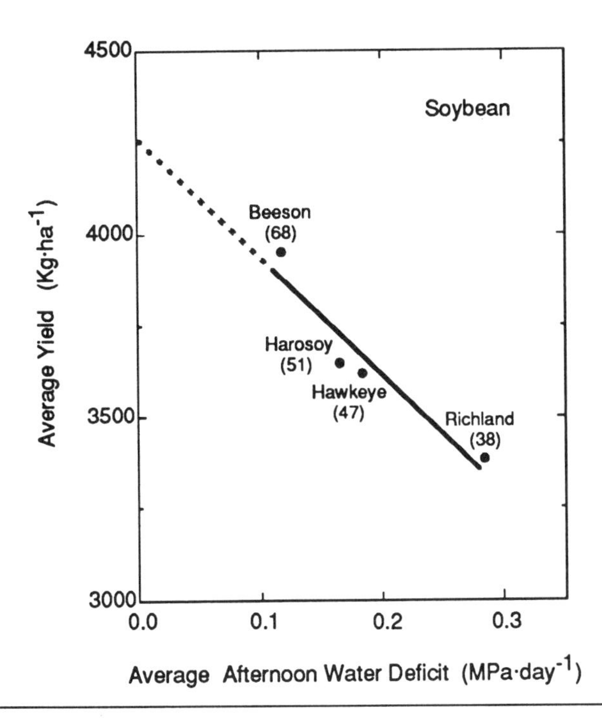

Figure 5.10. Average yields over three growing seasons and average afternoon water deficits in older and newer soybean cultivars measured in the field with a pressure chamber in Urbana, Illinois. The date of release of the cultivars is shown in parentheses. Older cultivars were part of the genetic lineage for the newest cultivar (Beeson). Water deficits were calculated as the average water potential below the threshold of -1.1 MPa each day. Extrapolating the dashed line to the Y-axis shows the yield increase expected if afternoon water deficits are eliminated. Water potentials were measured with a pressure chamber. Data from Boyer *et al.* (1980).

potential also increased. Thus, plant breeders were improving water transport while they improved yield. From such a plot, it can be predicted that the yield of Beeson can be improved another 9% if selections could bring midday water potentials to -1.1 MPa (extrapolation shown in Fig. 5.10). This process could be accelerated by measuring midday water potentials directly and selecting elite lines that have water potentials in the desired range.

There are certain measurement principles that need to be followed for successful cultivar comparisons in the field using a pressure chamber or psychrometer. First, careful attention is paid to keeping variation to a minimum. The main interest is in the midday water potential when photosynthesis is most rapid, and leaves are selected that are fully exposed to the sun and perpendicular to the incoming radiation. Second, the leaves are in the upper part of the canopy. This ensures that the flow path from the soil to the leaf is about the same length. Leaves lower in the canopy display water potentials that are less negative than in the upper canopy. Third, the comparisons are made during a part of the day when water potentials are stable. In the example of Fig. 5.9, this was done by repeatedly measuring a reference cultivar (Wayne) to evaluate the stability of the water potential. Comparisons were made between cultivars only during the stable time. Fourth, the pressure chamber is mounted on a portable cart and taken directly to the plant to be sampled or the psychrometer is taken to the field and loaded at the plant. This keeps sampling time to 10 sec or less and ensures that the water potential of the leaves is as reproducible as possible. Leaves should not be stored for later measurement.

Such precautions become essential because field comparisons are made in a variable environment that can cause the data to have too much scatter. The cultivar differences in Fig. 5.9 were about 0.2 MPa at the extreme and one would like to reliably detect even smaller differences. In the water potential range for soybean, small differences have a large effect on photosynthesis because the responsiveness of photosynthesis is so large (Boyer *et al.*, 1980).

Osmotic Adjustment

While comparing the water potential and osmotic potential of soybean seedlings, R.F. Meyer discovered that they adjusted osmotically when the root medium dehydrated (Meyer and Boyer, 1972). The adjustment was caused by an accumulation of solutes. As a result, the turgor scarcely changed. The maintenance of turgor implied that the water content of the tissue also remained high and this ultimately was shown to be the case (Bozarth *et al.*, 1987), indicating that the adjustment allowed more water extraction from the soil than otherwise would occur.

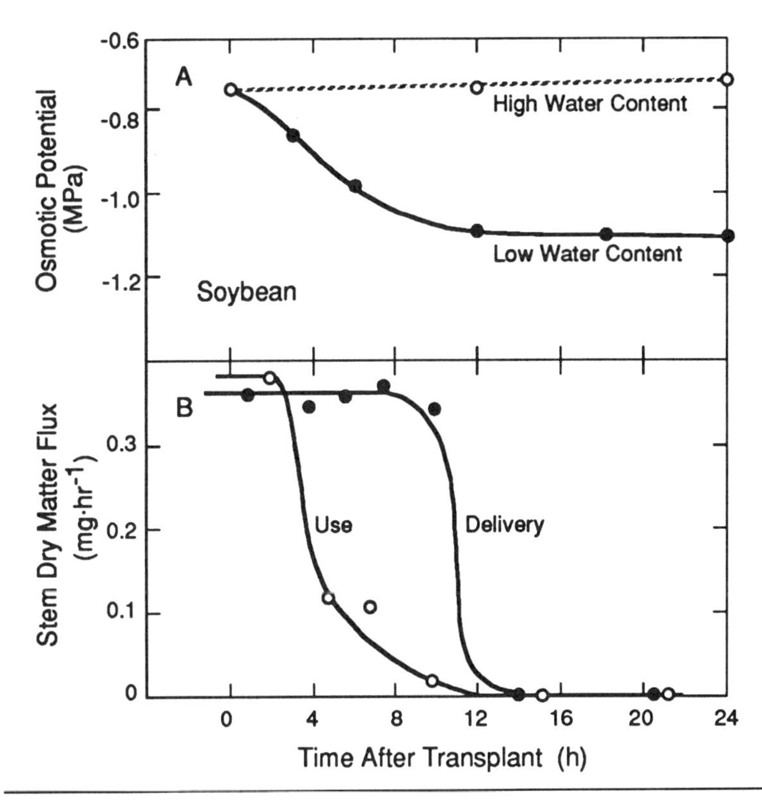

Figure 5.11. Osmotic adjustment and dry matter fluxes in the growing region of soybean stems after transplanting the seedlings to vermiculite of low water potential (Low Water Content) or high water potential (High Water Content). The decrease in osmotic potential (A) was caused by continued solute delivery to the growing cells but slower use (B). The osmotic potential became more negative when delivery exceeded use. Growth decreased in the first 2 hr after transplanting. The osmotic potential was measured with an isopiestic thermocouple psychrometer. Data from Meyer and Boyer (1981).

The solutes were mostly sugars and amino acids, and because dehydration did not occur in the tissues and the number of cells did not change in the growing region, there was an absolute increase in the number of moles of solute per cell (Meyer and Boyer, 1972). Figure 5.11A shows the decrease in osmotic potential as the root medium dehydrated, and Fig. 5.11B shows that solute delivery to the growing cells remained high but solute use declined (Meyer and Boyer, 1981).

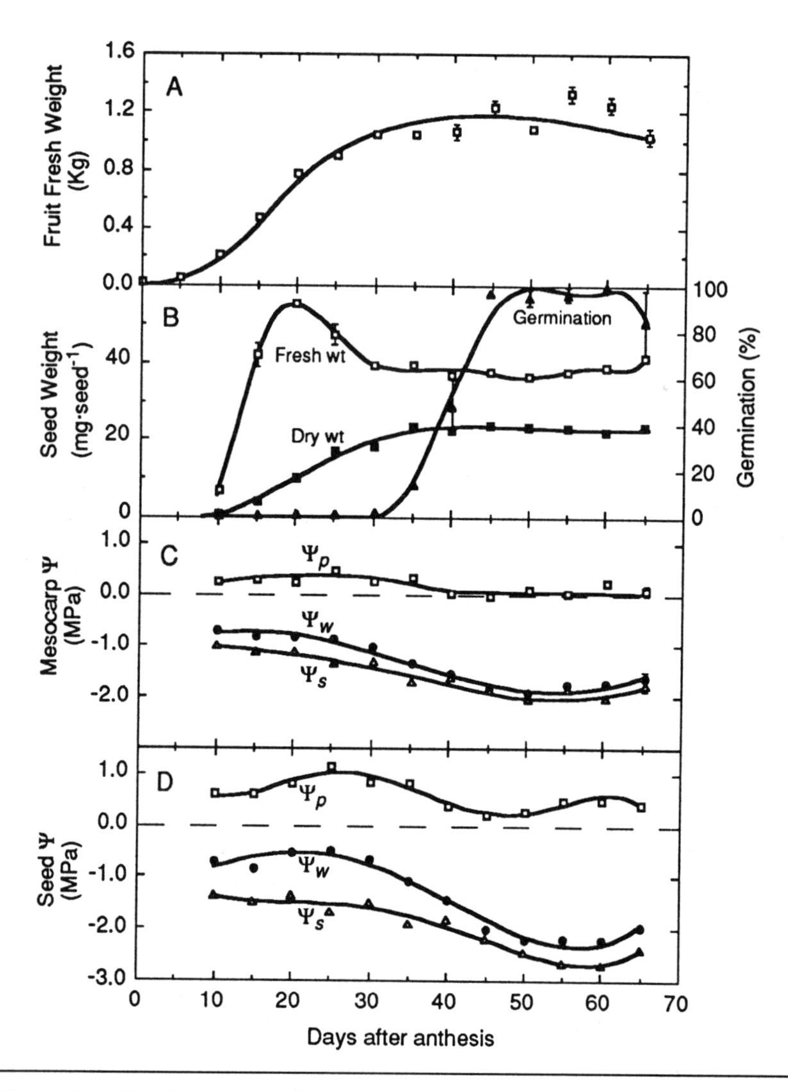

Figure 5.12. Development and water status at various times in muskmelon fruit and seed measured with a calibrated dew point hygrometer. A) Fruit fresh weight, B) seed fresh weight, dry weight, and germinability, C) water potential (Ψ_w), osmotic potential (Ψ_s), and turgor (Ψ_p) of the mesocarp of the fruit wall, and D) Ψ_w, Ψ_s, and Ψ_p of the seeds. Note the parallelism in Ψ_w, Ψ_s and Ψ_p between C and D. Data from Welbaum and Bradford (1988).

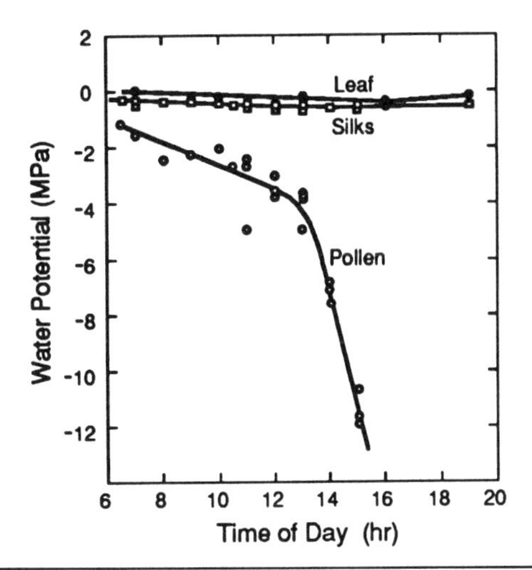

Figure 5.13. Water potentials of pollen, styles (silks), and leaves measured with an isopiestic thermocouple psychrometer in the same maize plants growing under field conditions in Urbana, Illinois. The pollen was freshly collected from anthers that had dehisced at dawn the same day. The silks were growing but were covered and had not been exposed to pollen. The leaves were sampled at the leaf tip which was mature. The psychrometers were rapidly loaded with samples in the field and transported back to the laboratory. Data from Westgate and Boyer (1986).

The high rate of delivery but low rate of use caused solute to accumulate in the cells resulting in the osmotic adjustment. After 15 to 20 hr, delivery also declined and came into balance with use (Fig. 5.11B). The balance caused the osmotic potential to stabilize, and no further changes occurred. Osmotic adjustment is thus driven by decreased solute use rather than increased transport or accelerated rates of hydrolysis.

Water Relations of Reproductive Tissues

The reproductive structures of plants develop rapidly and often require large amounts of water and solute. Welbaum and Bradford (1988) found that muskmelon fruits required about 30 days to enlarge during which the fresh weight gained over 1 kg most of which was water. The seeds enlarged early and laid down dry weight later (Fig. 5.12B), and throughout development the water potential of the mesocarp (fleshy wall of the fruit) and of the seeds was significantly below zero, suggesting that water potential gradients were present and favored water movement into the fruit (Figs. 5.12C and 5.12D). The water potential of the seeds decreased to -2.0 MPa at maturity (about 50 days) despite never having been exposed to a dehydrating atmosphere. The dehydration of the seed probably was caused by the low osmotic potentials surrounding the seeds (Figs. 5.12C and 5.12D).

Related work explored the water status of maize flowers and showed that the water potential of pollen was about -2.0 MPa at sunrise and -11 MPa by midafternoon (Westgate and Boyer, 1986), which was much lower than in growing styles (silks) or leaves (Fig. 5.13). This extreme desiccation occurred inside the locules of anthers that had dehisced at sunrise and illustrates that pollen desiccation is rapid even in that protected environment. The pollen was viable at the lowest water potential and able to fertilize the ovules. The low water potential allowed water to be absorbed from the surrounding stylar tissue and probably was essential for pollen tube growth. The styles were in contact with the water supply in the xylem and thus were able to supply water to the pollen tube.

References

Barrs, H. D. (1965). Comparison of water potentials in leaves as measured by two types of thermocouple psychrometer. *Aust. J. Biol. Sci.* **18,** 36-52.

Barrs, H. D. (1968). Determination of water deficits in plant tissues. *In* "Water Deficits and Plant Growth" (T.T. Kozlowski, ed.), Vol. 1, pp. 236-368. Academic Press, New York.

Blum, A., Sullivan, C. Y., and Eastin, J. D. (1973). On the pressure chamber technique for estimating leaf water potentials in sorghum. *Agron. J.* **65,** 337-338.

Boyer, J. S. (1966). Isopiestic technique: Measurement of accurate leaf water potentials. *Science* **154,** 1459-1460.

Boyer, J. S. (1967a). Leaf water potentials measured with a pressure chamber. *Plant Physiol.* **42,** 133-137.

Boyer, J. S. (1967b). Matric potentials of leaves. *Plant Physiol.* **42,** 213-217.

Boyer, J. S. (1968). Relationship of water potential to growth of leaves. *Plant Physiol.* **43,** 1056-1062.

Boyer, J. S. (1969a). Free-energy transfer in plants. *Science* **163,** 1219-1220.

Boyer, J. S. (1969b). Measurement of the water status of plants. *Annu. Rev. Plant Physiol.* **20,** 351-364.

Boyer, J. S. (1970). Differing sensitivity of photosynthesis to low water potentials in corn and soybean. *Plant Physiol.* **46,** 236-239.

Boyer, J. S. (1974). Water transport in plants: Mechanism of apparent changes in resistance during absorption. *Planta* **117,** 187-207.

Boyer, J. S. (1985). Water transport. *Annu. Rev. Plant Physiol.* **36,** 473-516.

Boyer, J. S. (1993). Temperature and growth-induced water potential. *Plant Cell Env.* **16,** 1099-1106.

Boyer, J. S., Cavalieri, A. J., and Schulze, E.-D. (1985). Control of cell enlargement: Effects of excision, wall relaxation, and growth-induced water potentials. *Planta* **163,** 527-543.

Boyer, J. S., Johnson, R. R., and Saupe, S. G. (1980). Afternoon water deficits and grain yields in old and new soybean cultivars. *Agron. J.* **72,** 981-986.

Boyer, J. S., and Knipling, E. B. (1965). Isopiestic technique for measuring leaf water potentials with a thermocouple psychrometer. *Proc. Natl. Acad. Sci. USA* **54,** 1044-1051.

Boyer, J. S., and Potter, J. R. (1973). Chloroplast response to low leaf water potentials. I. Role of turgor. *Plant Physiol.* **51,** 989-992.

Bozarth, C. S., Mullet, J. E., and Boyer, J. S. (1987). Cell wall proteins at low water potentials. *Plant Physiol.* **85,** 261-267.

Brown, R. W., and Collins, J. M. (1980). A screen-caged thermocouple psychrometer and calibration chamber for measurements of plant and soil water potential. *Agron. J.* **72,** 851-854.

Brown, R. W., and Oosterhuis, D. M. (1992). Measuring plant and soil water potentials with thermocouple psychrometers: Some concerns. *Agron. J.* **84,** 78-86.

Brown, R. W., and van Haveren, B. P., eds. (1972). "Psychrometry in Water Relations Research." Utah Agr. Exp. Sta., Utah State University, Logan, UT.

Campbell, E. C., Campbell, G. S., and Barlow, W. K. (1973). A dewpoint hygrometer for water potential measurement. *Agric. Meteorol.* **12,** 113-121.

Campbell, G. S., and Campbell, M. D. (1974). Evaluation of a thermocouple hygrometer for measuring leaf water potential *in situ*. *Agron. J.* **60,** 24-27.

Cavalieri, A. J., and Boyer, J. S. (1982). Water potentials induced by growth in soybean hypocotyls. *Plant Physiol.* **69,** 492-496.

Close, T. J., and Bray, E. A., eds. (1993). "Plant Responses to Cellular Dehydration During Environmental Stress." Amer. Soc. Plant Physiologists Series, Rockville, MD.

Cosgrove, D. J. (1985). Cell wall yield properties of growing tissue: Evaluation by *in vivo* stress relaxation. *Plant Physiol.* **78,** 347-356.

Cosgrove, D. J. (1987). Wall relaxation in growing stems: Comparison of four species and assessment of measurement techniques. *Planta* **171,** 266-278.

Cosgrove, D. J., van Volkenburgh, E., and Cleland, R. E. (1984). Stress relaxation of cell walls and the yield threshold for growth: Demonstration and measurement by micropressure probe and psychrometer techniques. *Planta* **162,** 46-54.

De Roo, H. C. (1969). Leaf water potentials of sorghum and corn, estimated with the pressure bomb. *Agron. J.* **61,** 969-970.

Dixon, H. H. (1914). "Transpiration and the Ascent of Sap in Plants." Macmillan, London.

Ehlig, C. F. (1962). Measurement of energy status of water in plants with a thermocouple psychrometer. *Plant Physiol.* **37,** 288-290.

Fellows, R. J., and Boyer, J. S. (1978). Altered ultrastructure of cells of sunflower leaves having low water potentials. *Protoplasma* **93,** 381-395.

Ghorashy, S. R., Pendleton, J. W., Peters, D. B., Boyer, J. S., and Beuerlein, J. E. (1971). Internal water stress and apparent photosynthesis with soybeans differing in pubescence. *Agron. J.* **63,** 674-676.

Gibbs, J. W. (1931). "The Collected Works of J. Willard Gibbs." Vol. 1. Longmans, Green and Co., New York.

Hanks, R. J., and Brown, R. W. (1987). "Proceedings of the International Conference on Measurement of Soil and Plant Water

Status." Vol. 1-3. Utah Agr. Exp. Sta., Utah State University, Logan, UT.

Harris, M. J., and Outlaw, W. H., Jr. (1991). Rapid adjustment of guard-cell abscisic acid levels to current leaf-water status. *Plant Physiol.* **95,** 171-173.

Hite, D. R. C., Outlaw, W. H., Jr., and Tarczynski, M. C. (1993). Elevated levels of both sucrose-phosphate synthase and sucrose synthase in *Vicia* guard cells indicate cell-specific carbohydrate interconversions. *Plant Physiol.* **101,** 1217-1221.

Huang, C. Y, Boyer, J. S., and Vanderhoef, L. N. (1975). Acetylene reduction (nitrogen fixation) and metabolic activities of soybean having various leaf and nodule water potentials. *Plant Physiol.* **56,** 222-227.

Hüsken, D., Steudle, E., and Zimmermann, U. (1978). Pressure probe technique for measuring water relations of cells in higher plants. *Plant Physiol.* **61,** 158-163.

Jachetta, J. J., Appleby, A. P., and Boersma, L. (1986). Use of the pressure vessel to measure concentrations of solutes in apoplastic and membrane-filtered symplastic sap in sunflower leaves. *Plant Physiol.* **82,** 995-999.

Kikuta, S. B., and Richter, H. (1988). Rapid osmotic adjustment in detached wheat leaves. *Ann. Bot. (London)* **62,** 167-172.

Klepper, B., and Kaufmann, M. R. (1966). Removal of salt from xylem sap by leaves and stems of guttating plants. *Plant Physiol.* **41,** 1743-1747.

Knipling, E. B. (1967). Comparison of the dye method with the thermocouple psychrometer for measuring leaf water potentials. *Plant Physiol.* **42,** 1315-1320.

Koide, R. (1985). The nature and location of variable hydraulic resistance in *Helianthus annuus* L. (sunflower). *J. Exp. Bot.* **36,** 1430-1440.

Kramer, P. J. (1985). An early discussion of cell water relations in thermodynamic terminology. *Plant Cell Env.* **8,** 171-172.

Kramer, P. J., and Boyer, J. S. (1995). "Water Relations of Plants and Soils." Academic Press, San Diego.

Malone, M., Leigh, R. A., and Tomos, A. D. (1989). Extraction and analysis of sap from individual wheat leaf cells: The effect of sampling speed on the osmotic pressure of extracted sap. *Plant Cell Env.* **12,** 919-926.

Malone, M., and Tomos, A. D. (1992). Measurement of gradients of water potential in elongating pea stem by pressure probe and picolitre osmometry. *J. Exp. Bot.* **43,** 1325-1331.

Maness, N. O., and McBee, G. G. (1986). Role of placental sac in endosperm carbohydrate import in sorghum caryopses. *Crop Sci.* **26,** 1201-1207.

Maruyama, S., and Boyer, J. S. (1994). Auxin action on growth in intact plants: Threshold turgor is regulated. *Planta* **193,** 44-50.

Matyssek, R., Maruyama, S., and Boyer, J. S. (1988). Rapid wall relaxation in elongating tissues. *Plant Physiol.* **86,** 1163-1167.

Matyssek, R., Maruyama, S., and Boyer, J. S. (1991b). Growth-induced water potentials may mobilize internal water for growth. *Plant Cell Env.* **14,** 917-923.

Matyssek, R., Tang, A.-C., and Boyer, J. S. (1991a). Plants can grow on internal water. *Plant Cell Env.* **14,** 925-930.

Meyer, R. F., and Boyer, J. S. (1972). Sensitivity of cell division and cell elongation to low water potentials in soybean hypocotyls. *Planta* **108,** 77-87.

Meyer, R. F., and Boyer, J. S. (1981). Osmoregulation, solute distribution, and growth in soybean seedlings having low water potentials. *Planta* **151,** 482-489.

Michel, B. E. (1972). Solute potentials of sucrose solutions. *Plant Physiol.* **50,** 196-198.

Molz, F. J., and Ferrier, J. M. (1982). Mathematical treatment of water movement in plant cells and tissue: A review. *Plant Cell Env.* **5,** 191-206.

Monteith, J. L., and Owen, P. C. (1958). A thermocouple method for measuring relative humidity in the range 95-100%. *J. Sci. Instrum.* **35,** 443-446.

Nelsen, C. E., Safir, G. R., and Hanson, A. D. (1978). Water potential in excised leaf tissue. A comparison of a commercial dew point hygrometer and a thermocouple psychrometer on soybean, wheat, and barley. *Plant Physiol.* **61,** 131-133.

Neumann, H. H., and Thurtell, G. W. (1972). A Peltier cooled thermocouple dewpoint hygrometer for *in situ* measurements of water potentials. *In* "Psychrometry in Water Relations Research" (R.W. Brown and B.P. van Haveren, eds.), pp. 103-112. Utah Agr. Exp. Sta., Utah State University, Logan, UT.

Nonami, H. (1986). "Mechanism of Cell Enlargement at Low Water Potentials." Ph.D. Dissert., University of Illinois, Urbana, IL.

Nonami, H., and Boyer, J. S. (1987). Origin of growth-induced water potential: Solute concentration is low in apoplast of enlarging tissues. *Plant Physiol.* **83,** 596-601.

Nonami, H., and Boyer, J. S. (1989). Turgor and growth at low water potentials. *Plant Physiol.* **89,** 798-804.

Nonami, H., and Boyer, J. S. (1990). Primary events regulating stem growth at low water potentials. *Plant Physiol.* **93,** 1601-1609.

Nonami, H., and Boyer, J. S. (1993). Direct demonstration of a growth-induced water potential gradient. *Plant Physiol.* **102,** 13-19.

Nonami, H., Boyer, J. S., and Steudle, E. S. (1987). Pressure probe and isopiestic psychrometer measure similar turgor. *Plant Physiol.* **83,** 592-595.

Nonami, H., Schulze, E.-D., and Ziegler, H. (1991). Mechanisms of stomatal movement in response to air humidity, irradiance and xylem water potential. *Planta* **183,** 57-64.

Outlaw, W. H., Jr., and Lowry, O. H. (1977). Organic acid and potassium accumulation in guard cells during stomatal opening. *Proc. Natl. Acad. Sci. USA* **74,** 4434-4438.

Passioura, J. B. (1980). The meaning of matric potential. *J. Exp. Bot.* **31,** 1161-1169.

Passioura, J. B. (1984). Hydraulic resistance of plants. I. Constant or variable? *Aust. J. Plant Physiol.* **11,** 333-339.

Passioura, J. B. (1988). Water transport in and to roots. *Annu. Rev. Plant Physiol. Plant Mol. Biol.* **39,** 245-265.

Pfeffer, W. (1900). "The Physiology of Plants." 2nd edition. English translation by A.J. Ewart. Oxford at the Clarendon Press.

Puritch, G. S., and Turner, J. A. (1973). Effects of pressure increase and release on temperature within a pressure chamber used to estimate plant water potential. *J. Exp. Bot.* **24,** 342-348.

Rawlins, S. L. (1964). Systematic error in leaf water potential measurements with a thermocouple psychrometer. *Science* **146,** 644-646.

Richards, L. A., and Ogata, G. (1958). Thermocouple for vapor pressure measurements in biological and soil systems at high humidity. *Science* **128,** 1089-1090.

Richter, H. (1978). A diagram for the description of water relations in plant cells and organs. *J. Exp. Bot.* **29,** 1197-1203.

Ritchie, G. A., and Hinckley, T. M. (1971). Evidence for error in pressure-bomb estimates of stem xylem potentials. *Ecology* **52,** 534-536.

Savage, M. J., and Cass, A. (1984). Measurement of water potential using *in situ* thermocouple hygrometers. *Adv. Agron.* **37,** 73-126.

Savage, M. J., Wiebe, H. H., and Cass, A. (1983). *In situ* field measurement of leaf water potential using thermocouple psychrometers. *Plant Physiol.* **73,** 609-613.

Savage, M. J., Wiebe, H. H., and Cass, A. (1984). Effect of cuticular abrasion on thermocouple psychrometric *in situ* measurement of leaf water potential. *J. Exp. Bot.* **35,** 36-42.

Scholander, P. F., Hammel, H. T., Bradstreet, E. D., and Hemmingsen, E. A. (1965). Sap pressure in vascular plants. *Science* **148,** 339-346.

Scholander, P. F., Hammel, H. T., Hemmingsen, E. A., and Bradstreet, E. D. (1964). Hydrostatic pressure and osmotic potential in leaves of mangroves and some other plants. *Proc. Natl. Acad. Sci. USA* **52,** 119-125.

Shackel, K. A. (1984). Theoretical and experimental errors for *in situ* measurements of plant water potential. *Plant Physiol.* **75,** 766-772.

Slatyer, R. O. (1967). "Plant-Water Relationships." Academic Press, New York.

Slatyer, R. O., and Taylor, S. A. (1960). Terminology in plant- and soil-water relations. *Nature* **187,** 922-924.

Spanner, D. C. (1951). The Peltier effect and its use in the measurement of suction pressure. *J. Exp. Bot.* **2,** 145-168.

Steudle, E. (1989). Water flow in plants and its coupling to other processes: An overview. *In* "Methods in Enzymology" (S. and B. Fleischer, eds.), Vol. 174, pp. 183-225. Academic Press, New York.

Steudle, E., and Zimmermann, U. (1971). Hydraulic conductivity of *Valonia utricularis*. *Z. Naturforsch.* **26b,** 1302-1311.

Steudle, E., Zimmermann, U., and Lüttge, U. (1977). Effect of turgor pressure and cell size on the wall elasticity of plant cells. *Plant Physiol.* **59,** 285-289.

Tomos, A. D., Steudle, E., Zimmermann, U., and Schulze, E.-D. (1981). Water relations of leaf epidermal cells of *Tradescantia virginiana*. *Plant Physiol.* **68,** 1135-1143.

Turner, N. C., and Long, M. J. (1980). Errors arising from rapid water loss in the measurement of leaf water potential by the pressure chamber technique. *Aust. J. Plant Physiol.* **7,** 527-537.

Tyerman, S. D., and Steudle, E. (1982). Comparison between osmotic and hydrostatic water flows in a higher plant cell: Determination of hydraulic conductivities and reflection coefficients in isolated epidermis of *Tradescantia virginiana*. *Aust. J. Plant Physiol.* **9,** 461-479.

Tyree, M. T. (1976). Negative turgor pressure in plant cells: Fact or fallacy? *Can. J. Bot.* **54,** 2738-2746.

Tyree, M. T., Cruiziat, P., Benis, M., LoGullo, M. A., and Salleo, S. (1981). The kinetics of rehydration of detached sunflower leaves from different initial water deficits. *Plant Cell Env.* **4,** 309-317.

Tyree, M. T., and Hammel, H. T. (1972). The measurement of the turgor pressure and the water relations of plants by the pressure-bomb technique. *J. Exp. Bot.* **23,** 267-282.

Tyree, M. T., and Jarvis, P. G. (1982). Water in tissues and cells. *In* "Encyclopedia of Plant Physiology" (O. Lange, P.S. Nobel, C.B. Osmond, and H. Ziegler, eds.), Vol. 12B, pp. 35-77. Springer-Verlag, Berlin.

Tyree, M. T., and Richter, H. (1981). Alternative methods of analysing water potential isotherms: Some cautions and clarifications. I. The impact of non-ideality and of some experimental errors. *J. Exp. Bot.* **32,** 643-653.

Tyree, M. T., and Richter, H. (1982). Alternate methods of analyzing water potential isotherms: Some cautions and clarifications. II. Curvilinearity in water potential isotherms. *Can. J. Bot.* **60,** 911-916.

Welbaum, G. E., and Bradford, K. J. (1988). Water relations of seed development and germination in muskmelon (*Cucumis melo* L.). *Plant Physiol.* **86,** 406-411.

Westgate, M. E., and Boyer, J. S. (1984). Transpiration- and growth-induced water potentials in maize. *Plant Physiol.* **74,** 882-889.

Westgate, M. E., and Boyer, J. S. (1985). Osmotic adjustment and the sensitivity of leaf, root, stem, and silk growth to low water potentials in maize. *Planta* **164,** 540-549.

Westgate, M. E., and Boyer, J. S. (1986). Silk and pollen water potentials in maize. *Crop Sci.* **26,** 947-951.

Zhu, G. L., and Boyer, J. S. (1992). Enlargement in *Chara* studied with a turgor clamp: Growth rate is not determined by turgor. *Plant Physiol.* **100,** 2071-2080.

Zimmermann, U., and Steudle, E. (1978). Physical aspects of water relations of plant cells. *Adv. Bot. Res.* **6,** 45-117.

Zimmermann, U., Hüsken, D., and Schulze, E.-D. (1980) Direct turgor pressure measurements in individual leaf cells of *Tradescantia virginiana*. *Planta* **149,** 445-453.

Index

M

N

O

P